Afterwards

Folk and Fairy Tales
with Mathematical Ever Afters

Afterwards

Folk and Fairy Tales
with Mathematical Ever Afters

Peggy Kaye

Cuisenaire Company of America, Inc.
White Plains, New York

Managing Editor: Alan MacDonell
Acquisitions Editor: Doris Hirschhorn
Developmental Editor: Harriet Slonim

Design Manager: Phyllis Aycock
Cover design and illustration: Tracey Munz
Text design, line art, and production: Fiona Santoianni
Illustrations: Peggy Kaye

2 3 4 5 VG 01 00 99 98

Table of Contents

About the Author

Peggy Kaye is the writer and illustrator of both volumes of *Afterwards*—grades 1 and 2 and grades 3 and 4. She is also the author and illustrator of the *Games For* series: *Games For Math, Games For Reading, Games For Writing,* and *Games For Learning.* These books give parents and teachers hundreds of playful ways to help children master reading, writing, and arithmetic.

She is also the author of the *Homework: Math* and the *Homework: Reading* series. These first- through sixth-grade workbooks encourage parent participation in their children's education. These books are now available in bookstores under the names *Enrichment Math* and *Enrichment Reading.*

In addition to books, she has written numerous articles and reviews for a wide range of magazines and newspapers.

Peggy Kaye lives in New York City where she writes, draws, tutors children, and works as an educational consultant.

Afterwards

Once upon a time, a teacher began telling fairy tales to her students. Every Friday afternoon she sat on a pillow-strewn floor, surrounded by eager youngsters, and began her story. Her students loved this Friday ritual. Even the roughest, toughest, most street-savvy kids fought to sit on the teacher's lap during tale time. And so the teacher discovered what she probably should have known all along—that fairy tales and folk tales touch something deep and abiding in children.

Years later, the teacher began reading about innovative educators who were using literature as a springboard for math lessons. Now, this teacher thought, why not hook up fairy tales and math? And so that is just what the teacher did.

Here is the result: a collection of nine tales and related math activities. The stories come from all over the world. The collection includes a story from *1001 Arabian Nights*, an Iroquois tale, and an African-American legend. There are tales from Japan, Germany, Ireland, Ghana, and Russia along with a Jewish tale first told in Iraqi Kurdistan. The collection includes trickster tales, humorous stories, and yarns of love, sacrifice, and bravery. In the stories you'll meet an evil magician, a wicked witch, a flying horse, and a clever giantess. In other words, there is something here for everyone.

Why is the book called *Afterwards*? Have you ever wondered what happens after "happily ever after"? If the people in a story are vivid enough, interesting enough, compelling enough, children delight in contemplating the characters' post-story lives. That's what they get with an *Afterwards* math story: an epilogue with a mathematical twist to a classic tale.

Imagine this. In the course of the Russian story, *Baba Yaga*, a girl's kind heart and generous nature help her escape from an evil witch. What will she do after the tale ends? Well, she decides to make a spell repellent so that she never has to worry about that witch again. The repellent must have exactly 100 centiliters of liquid made up of at least six different ingredients. Can your students help her pick just the right amount of each liquid so that she has a perfect repellent? That's a math problem.

The girl decides to make a house for a little mouse who helped her avoid becoming the witch's evening meal. The house will have five rooms and cover an area of 225 square feet. Your students' job is to design the rooms and discover the area and perimeter of each one. That's another math problem.

In crucial ways, *Afterwards* word problems differ from their basal cousins. As you will discover, the problems tend to be more thought-provoking. They may take longer to solve. They demand complex thinking skills. But the most important difference is, perhaps, a bit more subtle. Having heard the folk or fairy tale, children know the characters in *Afterwards* stories and have an authentic interest in their concerns, pleasures, and problems. And so these problems excite, rather than deaden, children's imaginations.

Using the Tales

There are nine tales in *Afterwards*. Each story is followed by several math activities. There is no established order to the stories. Choose any one that seems appealing and begin. The same holds true for the math activities. Pick the one(s) you think are most appropriate for your students and ignore the rest. In general, it will take about 10 minutes to read the tale and another 20 to 30 minutes for children to solve each problem. You may decide to present more than one activity for any given tale. Do this on consecutive days or assign different activities to different groups of students.

Presenting the Tales and the Activities

Your first step is to share the stories with your class. You can read them aloud, of course, but you might consider telling a story or two without the text. Something special seems to happen when you tell, rather than read, a tale—maybe because these stories are rooted in an oral tradition. But it does take a little bit of work to master a tale before telling it. If you do want to put in the effort, don't try to memorize the tale. In general, storytellers prefer using their own words to create a spontaneous texture and tone. You can even transform the tale as you are talking, adding to the plot or leaving something out. Believe me, the Grimm brothers' ghosts will stay quiet in their graves. Some of the more fluent readers in your class can, no doubt, read the stories independently. All you need to do is provide them with photocopies.

Once you've presented a tale, turn to the teacher pages for the math activities. Select one that is suitable for your class. Then read it aloud or explain it in your own words. Using your own words probably makes the most sense when the activity instructions are complicated. Feel free to make any changes you want in the story or in the mathematical content. Use different numbers. Make the problem easier or harder. Extend it. Condense it. Vary it. Whatever you do to it is perfectly all right.

Most of the activities also appear on reproducible student pages. If you don't want to photocopy these pages, you can always write the needed information on the chalkboard or on a sheet of chart paper.

Grouping

Is it better to present the activities to your entire class or to smaller math groups? Should children work independently or in teams? There are no rules. The choice is yours. And you needn't feel tied down to one approach. Some problems will lend themselves more naturally to whole-class collaborations—others to small-group work or to individual effort.

You do not need to present an activity immediately after reading a story. You can, for example, share a tale during a class meeting on Monday, present one activity to a math group on Tuesday and another activity to a different group on Wednesday.

What should you do when a child or team of children finishes before the rest of your class? There are several ways to handle this situation. You could ask each child to write a short explanation of his or her work. Alternatively, give the child or children a new *Afterwards* activity. Better yet, encourage children to make up their own problems based on the *Afterwards* tale.

About the Activities

I've tried to include a variety of activities in *Afterwards*. There are addition, subtraction, multiplication, and division problems; fraction problems, probability problems, and problems involving measurement, money, and estimation. The problems call on children to think logically, organize information, and search for patterns and they employ various problem-solving strategies such as trial-and-error and working backwards. As in life, some problems have only one answer; others have several possible solutions. The difficulty varies, too. This should make it easy for you to find just the right assignment for your students. All the problems, however, have one thing in common—they demand real thought and effort to solve.

What is a problem? Is finding the sum of 8 and 7 a problem? For you? Not really. It is more accurate to describe it as an unanswered question. A true problem demands concentrated intellectual energy to solve. It's never easy to unravel an authentic problem, and there is always a possibility of failure, no matter how hard you try. Proving Fermat's last theorem; well, now, that's a problem.

Afterwards is full of activities that third and fourth graders should find challenging (perhaps not as awesomely so as validating Fermat's last theorem, but challenging nevertheless). The activities will require time and effort to finish. At first, some children might have a hard time doing the necessary work. They may need considerable support from you. You must assure them that speed and praiseworthy work are not synonymous. Applaud youngsters' willingness to spend time exploring various ways to solve the problem. As children gain experience, the problem-solving process will become more natural. Children will feel confident and at ease. They will discover that intellectual intensity has a pleasure all its own.

Beginning an Activity

After presenting an activity to your class or math group, spend some time discussing it. Make sure everyone understands what the problem asks. You may call on your students to restate the problem in their own words. Then elicit suggestions for solving the problem. If students fumble for ideas, offer one or two. Don't present a total strategy, just nudge your pupils in a promising direction. It helps if you ask questions such as, "Do you think we should draw a picture of Spider's webs?" Or, "Would making a chart help help you keep track of the seating arrangements at Lubachka's table?" Or, "Why not try out a set of numbers and see what happens?"

Then let the children take over. Don't interfere if your students come up with less-than-promising proposals. Soon enough they will discover the pitfalls in their approaches, and then they will have to rethink. If they are truly stymied, feel free to help out. You can also encourage children to discuss their work with their classmates in order to discover how they are proceeding.

Good Strategies

When it comes to understanding a problem, nothing is more powerful than recreating the story for children. There are three excellent ways to do this.

- First, children can draw pictures. In "Everyone Wants to Fly," one of the activities that follow *The Ebony Horse*, children must determine the time it takes for fourteen riders to each get a turn flying on a magic horse. By drawing pictures of the riders, children will find it considerably easier to keep track of, and record, the passing time.

- Second, children can act out the problem. This would certainly be a good alternative approach to "Everyone Wants to Fly." Fourteen children can portray members of the royal family and a fifteenth child can be the palace timekeeper. The timekeeper, holding a clock with movable hands, can keep track of the minutes, hours, and days as each rider takes his or her turn on the horse.

- Third, your students can recreate the story using manipulative materials. In one activity, children will try to design lettuce gardens for

Rapunzel. As they do so, they can make excellent use of 1-cubic-inch blocks. When building a house for Skywoman, the featured character in *On Turtle's Back*, children can use color cubes to create a model of her new home.

Some third and fourth graders resist using these techniques. They may worry that it is "babyish" to draw pictures, play act, and manipulate counting materials. But if you demonstrate—draw a bit, direct a short dramatization, or pull out base-10 blocks—you can prove how useful it is to create a concrete model of the problem.

The activities in *Afterwards* offer a perfect opportunity to introduce your students to several important problem-solving strategies, such as trial-and-error, working backwards, searching for patterns, working with smaller numbers, and making an organized list. When it is appropriate, suggest a strategy, or, better yet, initiate one. You do not need to complete the activity for the children—you can simply get things going.

In "Oonagh's Necklace," an activity that follows *The Legend of Knockmany*, students must determine the number of copper, brass, silver, and gold links in Fin's collection. Get children to suggest a number, any likely number, to tell how many copper links Fin has and then have them use that number to work out the problem. Are the results too large? Then your students must select a second number of copper links—a smaller number, of course, than the first. Are they closer to the right answer now? Once they have the idea, they can continue working on the problem independently until they find the proper combination of links.

As you work through the book, you will begin to see changes in how your students approach the activities. In a surprisingly short time, students may begin employing sophisticated problem-solving strategies even without your prompting. When you expect great things from children, they meet the challenge.

Crossing the Curriculum

There are a variety of ways to link *Afterwards* stories to your language arts, social studies, and science programs. Here are a few thoughts.

Language Arts
- Children can write their own versions of a particular *Afterwards* tale. They can then illustrate their stories and "publish" them.
- Children might also write about what happens to the characters after a story ends. Their work does not need to have a mathematical "hook."

- Encourage children to make up their own fairy tales based on the *Afterwards* stories.
- Read more tales to your class. There are so many good ones, you could easily make such storyreading a daily event.
- Let children be storytellers, sharing their favorite tales with the class.
- Groups of children can put on a play based on their favorite tale. They can improvise scripts or they can write scripts and then memorize their own parts.

Social Studies
- Get a world map and stick it with pins indicating each story's country of origin.
- Cook ethnic foods from a selected *Afterwards* country.
- Find out about the music and art of the country.
- Do mini-research projects on *Afterwards* countries.

Science
- Several of the stories feature animals. Have children research the life cycles of spiders, turtles, hound dogs, cranes, or any of the other creatures in the *Afterwards* stories.

Of course, it isn't necessary to extend the tales. This is your program now, and you should feel free to use it in any manner you desire. In that way, I hope you will enjoy your *Afterwards* math lessons and that they all end happily ever after.

Dinner with Spider and Turtle

A Tale from Ghana

Spider was always hungry. It seemed the more he ate the more he wanted to eat. He consumed everything in sight and never willingly shared his food. At mealtimes, all the villagers knew to stay far away from Spider.

But one evening, a stranger—Turtle by name—came to the village. Turtle, who had been traveling since dawn, was tired, hot, and hungry. So he knocked on the first door he saw. It was Spider's door. Of course, Turtle expected a warm reception. After all, everyone knows how important it is to show proper hospitality to a stranger. Even Spider knew the rules of hospitality. If he turned his back on a stranger, the villagers would find out and say nasty, oh very nasty, things about him.

So Spider did welcome Turtle. And, since it was dinnertime, he offered Turtle something to eat. Of course, Spider hated parting with his food. But hospitality is hospitality, so what could he do? Well, actually, he did think of something.

Before Turtle stepped through the doorway, Spider said, "My friend, you are exceedingly dusty. Why don't you walk to the stream and wash your feet? That way you will not mess up my dining room. Go on now, follow the path. You'll get to the stream soon. Meanwhile, I'll prepare dinner."

Looking at his feet, Turtle replied, "You are right, Spider. I certainly will not disgrace your home by entering with filthy feet."

That said, Turtle hiked down the dirt road to the stream. He washed his feet until they glistened. Then he headed back up the dirt road to Spider's house. Naturally, his feet were grimy again before he reached Spider's door.

As Turtle approached the house he saw, through the window, platters of food covering the kitchen table. Everything looked delicious. Turtle could hardly wait to sit down and enjoy the meal.

But when Spider opened the door, he cried, "Turtle, look at your feet! They are covered with dust and mud. Would you mind going to the stream and washing them before coming into my home?"

Turtle admitted that his feet were dirty, and so he trudged back to the stream. There he scrubbed and scrubbed until his feet were sparkling clean.

By the time Turtle returned to the house, his host was already eating. Indeed, Spider was gobbling food. Did he invite Turtle to join him? No, he did not. Instead he said, "Dear friend, isn't this a marvelous meal? You must agree, it is the best you've ever known. But look here, Turtle—your feet—they are filthy. Please, do not track mud all over my spotless floor. Go to the stream and bathe yourself."

"I've washed and washed," said Turtle, "but your road is extremely dusty, and I cannot keep my feet clean."

"Really, Turtle," said Spider as he took another mouthful of baked yams, "here I am sharing this lovely meal with you, and yet you insult my road. But I forgive you. Just clean yourself and together we will finish dinner."

And so Turtle walked to the stream. This time though, after washing his feet, he wrapped grass around each one. Again he returned to the house. But now his grass boots kept him from getting dirty.

When Turtle arrived, however, the kitchen table was empty. There was not a nibble, a morsel, or a crumb of food in sight. What did Spider have to say? Just this: "Excellent dinner, don't you agree, Turtle. It surely was an excellent dinner."

Turtle did not argue. Instead he smiled and said, "Yes, Spider, you certainly know how to treat a guest. If you are ever in my village, please visit me for a meal." Without saying another word, Turtle left Spider's house.

Some months later, Spider did visit Turtle's village. He went to the lake where Turtle made his home and found him sleeping by the shore.

"Welcome old friend," said Turtle when he saw Spider. "How wonderful to see you. You will join me for dinner, I hope."

"Why, yes," said Spider, "I accept your hospitality as you accepted mine."

"Excellent, excellent," said Turtle. "Wait here while I attend to our food."

Turtle dove into the water and swam to the bottom of the lake. There he prepared a glorious feast. After setting out the meal, he returned to the surface and announced, "Dinner is ready, Spider. Come, let's eat." Then Turtle lowered himself under the water again.

Spider wanted to follow, but he was such a lightweight creature, he could not get under the surface. Over and over he tried to dive down, but it did no good. He kept popping back up. Worst of all, Spider could see through the clear water all the way to the bottom of the lake. There was Turtle enjoying a scrumptious-looking meal.

After a few minutes, Turtle swam to the surface and asked, "Spider, why don't you join me? The food is delightful. Come on, hurry up."

Then, before Spider could say anything, Turtle dove back to the bottom of the lake.

What could Spider do? How could he get to the food? He thought and thought, and then he had an idea—a very clever idea. He made his way to shore. There he collected lots and lots of pebbles. He stuffed them into the pockets of his jacket. With his pebbles in place, Spider was no longer a lightweight. He was a heavyweight. He dove into the water and, this time, swam straight down.

Finally, Spider reached the dinner table. He was hungry, so hungry. And the food looked good, so good. But before he could take even one mouthful, Turtle said, "Really Spider, you must not eat with your jacket on. It is not polite." With these words, Turtle removed Spider's pebble-filled jacket. Instantly, Spider drifted away from the table, away from the food, and up to the top of the lake. From the surface, he looked down and saw Turtle feasting on the last bites of dinner.

And now you know why people say, "*One good meal deserves another.*"

Notes

About the Story

Dinner with Spider and Turtle bears a striking resemblance to Aesop's fable *The Fox and the Crane*. Both tales involve meals that cannot be consumed by guests and both end with a moral. It might be fun to read Aesop's fable to your class after reading *Dinner with Spider and Turtle*. Then your students can discuss the similarities and the differences between the two stories. Students may then want to write their own stories in which the moral is: *One good meal deserves another.*

For this retelling, I relied on one source.

> Courlander, Harold and Herzog, George. *The Cow-Tail Switch and Other West African Stories*. New York: Henry Holt and Company, 1974, 107-112.

Your Thoughts

Dinner with Spider and Turtle

Spider's Dinner

This counting problem focuses on multiples of four. Students must identify the pattern 4, 8, 12,... and continue it for a total of eight fours, or 32. Next, they must add the numbers that form the pattern. When they do, they get a sum of 144. That's the number of bugs in the webs. Finally, they must subtract 144 from 150 to find how many more bugs Spider needs for his pie. Once they realize that he needs 6 more bugs, students can suggest how he can go about getting them.

When he arrives home from Turtle's lake, Spider is extremely hungry. He decides to make a scrumptious bug pie. To prepare this delight, he needs 150 fresh bugs.

Now it so happens that Spider has eight excellent insect-catching webs in his house. Checking them, he discovers fat juicy bugs in one and all. Surprisingly, there is a pattern to the number of bugs in his webs. The first web has 4 bugs. The next has 8. The next 12.

Continue the pattern to find out how many bugs are in each of the webs.

If Spider gathers the bugs in all eight webs, will he have enough bugs for his pie? If not, what could he do?

Dinner with Spider and Turtle

Spider's Dinner

When Spider gets home, he decides to make a bug pie. To prepare it, he needs 150 bugs.

Spider looks at his eight webs and discovers a surprising pattern. The first web has 4 bugs. The second web has 8 bugs. The third has 12.

Go ahead and continue the pattern.

If Spider uses all these bugs,
will he have enough for his pie?
If not, what can he do?

Dinner with Spider and Turtle

Spider Facts

After students solve and discuss both word problems presented here, they can use the list of spider facts to invent their own problems. Consider collecting their work and compiling it into a class anthology of spider problems.

Spiders are not insects. They are arachnids. A spider has eight legs and its body is in two parts. An insect has six legs and three body parts.

Here is a list of more facts about spiders. Read the facts and then use them to make up your own math problems. Here are a couple of problems to get you started.

- Imagine you are facing 8 spiders. How many eyes are staring back at you?

- How many leaps would it take a jumping spider to cross our classroom?

Dinner with Spider and Turtle
Spider Facts

Here is a list of facts about spiders. Use the facts to make up math problems.

- There are more than 30,000 different species of spiders in the world.

- All spiders have 4 pairs of legs.

- Most types of spiders have 4 pairs of eyes.

- Tarantulas can live much longer than other spiders. Females can live for 20 years. Males usually live only for 10 years. The oldest tarantula on record lived for 28 years.

- Tarantulas are the largest spiders. Most tarantulas living in the United States grow to a length of between 2.5 cm and 7.5 cm. Some tarantulas living in other places grow a lot bigger. One kind, the goliath bird-eating spider of South America, is so big it can eat both birds and rats. The largest female goliath on record was 10 cm long, had a leg span of 26.3 cm, and weighed 123.2 g.

- The smallest type of spider is found in Western Samoa. It measures 0.5 mm. That is smaller than the period at the end of this sentence.

- One type of spider in New Guinea can weave a web that is 2.4 m across. Fishermen use these webs as fishing nets.

- Most spiders travel at a rate of about 1.8 km/h. Some gecko- and lizard-eating spiders in Africa and the Middle East can reach speeds of more than 16 km/h.

- Some spiders travel by floating through air. They can go as far as 600 m during a single floating session. Over time they can travel more than 300 km.

- A jumping spider can cover a distance of 20 cm in one leap.

- A female spider's egg sac can contain as many as 3,000 eggs.

Dinner with Spider and Turtle

Turtle Facts

Some students may prefer to make up turtle problems rather than spider problems. Others may want to combine turtle and spider facts into single problems.

Turtles vary enormously in size and living habits. Here is a list of facts about turtles. Read the facts and then use them to make up your own math problems. Here are a few problems to get you started.

- If a box turtle is born in the wild today and lives for 23 years, in what year will he die?

- How many bog turtles lined up nose to tail would it take to equal the length of a leatherback turtle?

- If a green turtle reaches her nesting ground on February 15 and stays until June 15, how many times will she lay eggs?

Dinner with Spider and Turtle
Turtle Facts

Here is a list of facts about turtles. Use the facts to make up math problems.

- There are about 250 species of turtles in the world. Of these, about 200 species live in water. These turtles come onto land only to lay eggs.

- The bog turtles are the world's smallest turtles. They grow to a length of about 7 cm.

- The largest sea turtle, the leatherback, can grow to more than 2 m and weigh more than 500 kg.

- The smallest sea turtle is the very rare Atlantic ridley. It grows to a length of 63.5 cm and can weigh 36 kg.

- The largest freshwater turtle is the alligator snapping turtle. It can grow to 76 cm and can weigh 107 kg.

- In captivity, some turtles have lived more than 100 years.

- In nature, box turtles and slider turtles can live 20 to 30 years.

- On land, most turtles can travel at a rate of about 0.25 km/h.

- Leatherback turtles can swim at a speed of 32 km/h.

- A female sea turtle lays 150 or more eggs at a time. Sea turtle eggs take 5 to 10 weeks to hatch. Each egg is about 50 mm long.

- Female green turtles swim more than 1,600 km to reach their nesting grounds. They arrive each February and stay until June. While there, they lay eggs every 12 days—about 100 eggs at a time.

Dinner with Spider and Turtle
Cake for Everyone

The students' job here is to determine the number of large and small cakes the baker must prepare. Each small cake can be cut into 10 slices; each large cake into 15. To advise the baker, children must add a multiple of 10 and a multiple of 15 in order to get a sum of 300.

The problem can be solved using addition and subtraction alone, although some students may also use multiplication. Students will probably use trial-and-error until they find a correct combination of large and small cakes. There are many correct solutions including 8 large cakes and 18 small ones, 10 large cakes and 15 small ones, and 18 large cakes and 3 small ones.

In Spider's village, everyone was delighted to hear how Turtle tricked their selfish, greedy neighbor. The villagers were so happy, they decided to have a party.

For this event, the village baker will prepare cakes. The baker must have 300 slices to feed all the villagers. He will bake large cakes and cut them into 15 slices each. He will bake small cakes and cut them into 10 slices each.

The baker has a problem, though. He does not know how many large and small cakes to prepare. Can you advise him?

Dinner with Spider and Turtle

Cake for Everyone

There is going to be a party.

Everyone in Spider's village will be there.

They are celebrating because clever Turtle managed to trick selfish Spider.

The baker will prepare cakes.

He needs 300 slices for the villagers.

He will bake large cakes and small ones.

Each large cake can be cut into 15 slices.

Each small cake can be cut into 10 slices.

How many large cakes and small cakes should the baker bake?

Dinner with Spider and Turtle
Spider's Game

Students may think that Turtle has the better deal in this game because he appears to have six ways of getting a point, while Spider appears to have only five. Yet, after playing the game and comparing their results, students will discover that Spider is the one more likely to win. This is because there are more possible dice combinations for Spider's numbers than for Turtle's. As children try to explain Spider's advantage, they get a lesson in probability and mathematical reasoning.

Sum of Numbers Rolled	2	3	4	5	6	7	8	9	10	11	12
Number of Combinations	1	2	3	4	5	6	5	4	3	2	1

One day Spider and Turtle happened to meet in a small village halfway between their homes. Immediately, Spider apologized for tricking Turtle and Turtle apologized for his own trick.

Then Spider made a suggestion, "Since we are so close to your house, why don't you invite me to dinner like a good friend?"

Turtle answered, "Actually, we are just as close to your house and I think you should invite me to dinner—to the one I missed."

"I'll tell you what," said Spider. "Let's play a game. The loser will prepare dinner for the winner."

Turtle agreed, and Spider explained these game rules.

"We will take turns rolling a pair of dice. On each turn, we will add the two numbers rolled. If the sum is 2, 3, 4, 10, 11, or 12, you—Turtle—will get one point. If the sum is 5, 6, 7, 8, or 9, I—Spider—will get one point. We will each roll 20 times. Whoever gets the most points wins. Whoever loses has to prepare dinner."

Do you think Spider's game is fair? Explain your answer.

Play the game with one or more friends. Give one player (or team) Turtle's winning numbers. Give the other player (or team) Spider's.

Who won? Compare your results with other players or teams.

Now do you think the game is fair? Explain your answer.

The Crane

A Tale from Japan

Once upon a time an old man and an old woman lived in a small, but comfortable cottage near a dense marsh. Every day the man wandered through the marsh gathering plants. Every night his wife cooked the plants for their meal.

One afternoon, as the man worked, he heard an odd cry coming from the tall, thick reeds. It was such a strange and sad noise, the man felt compelled to investigate. What did he see? A magnificent white crane caught in a trap. The large bird was struggling, fighting, straining to free herself. But the harder she fought, the more tightly she became ensnared. The animal was frantic.

The man, overcome with pity for the great bird, decided to set her free. Slowly, he walked toward the crane, whispering reassuring words. His voice was so gentle, the bird stopped struggling. Carefully, tenderly, the old man removed the trap from the bird's leg. After gaining her freedom, the crane stared at the man with grateful eyes. Then she rose into the air and started flying.

The sun reflected silver light across the bird's enormous white wings as she circled once, twice, three times over her rescuer's head. The old man had never seen such a magnificent sight.

That night, while eating dinner, the man told his wife about his adventures with the crane. "Oh, how I ache for that poor, frightened bird," said the wife. "It is fortunate that you found her and set her free."

Very early the next morning, the old couple heard someone knocking on their cottage door. Opening the door, they were greeted by a beautiful girl. It seemed the girl had been traveling for a long time and was both tired and hungry. The old woman prepared a hot meal for her guest. As she ate, the girl revealed that she had no family, no home, nowhere to live, nowhere to go.

Immediately the old woman implored, "Please stay with us. Be the daughter my husband and I have yearned for all of our lives. We cannot give you fine clothes or elegant meals, but we will give you all our love."

Happily, gratefully, the girl accepted the invitation.

As the weeks passed, the old man and the old woman discovered that their new daughter was kind and generous. Not only that, she was a hard worker. She gathered plants in the marsh with the old man, and somehow she always managed to find the tastiest ones. She helped the old woman with cooking and cleaning. She was always cheerful, always smiling, and she brought unimagined joy to her new parents.

One morning, the girl said, "Mother, in my room, there is a loom. May I use it to weave sometimes?"

"Of course," said the old woman.

"When I weave," said the girl, "I must be completely alone. I beg you, never enter my room while I work. Please, do not even peek inside."

Although puzzled by this request, the old woman promised to do as her daughter asked.

After breakfast the next morning the girl began weaving. All morning the old couple could hear the loom going *clickety clack, clickety clack* through the thin walls. The girl did not leave her room at lunchtime. She did not leave to come to dinner.

While eating their evening meal, the old man and woman worried about their daughter. Why did she stay so long in her room? Could she be sick or hurt? Should they open her door and check? But they had promised not to disturb her. What could they do? What should they do?

Before they could decide, the girl joined them. She looked tired—so tired—and pale—so pale. Slowly she walked to her parents and handed them her weaving. It was the most spectacular woven cloth that the old couple had ever seen. It showed the marsh at sunrise. Golden rays filtered through the air and caressed swaying reeds. The pink-and-pale-blue sky was full of cranes, their wings sparkling with light.

"Mother, Father," said the girl, "take this cloth to the marketplace tomorrow and sell it. Someone will pay well for it, I am sure."

The old couple wanted to keep the weaving; it was so beautiful. But the girl insisted that they sell it.

The next morning the old man went to the marketplace. Everyone there wanted to own the magnificent cloth. One after another people offered the old man money and more money. Finally, one wealthy gentleman said he would pay three ryo of gold—an enormous sum! And so the old man sold the weaving.

Before returning home, he bought chicken for dinner, pears and sweet ginger for after-dinner treats, and new kimonos for his wife and daughter. Even after purchasing these many delights, a considerable amount of money remained in the man's pockets.

That night the old couple and their daughter had a wondrous feast. For several months afterwards the family enjoyed unheard of luxuries—fish or meat every day, fruit and sweets, new sandals, and jade hair combs.

Finally, though, the money ran out. When that happened, the girl returned to her loom. The old couple heard *clickety clack* all morning, *clickety clack* all afternoon, *clickety clack* into the night. At last, after hours and hours, the girl came out of her room. The weaving she handed her parents this time was even more glorious than her first.

Night was there on the cloth. And it was the most beautiful night imaginable. Sparkling silver stars and a luminous crescent moon dotted the purple-black sky. The stars and moon appeared again—as a reflection on a shimmering lake. By the lake, next to a row of tall rushes, stood a lone crane. Her tilted head faced forward. This woven bird had such a lively and intelligent expression that, at first glance, she seemed to be alive.

The old couple wanted to keep the weaving, but the girl insisted that they sell it. And so the man took the cloth to the market-place. There, a wealthy gentleman paid six ryo of gold for it.

For many, many months, the old couple and the girl had all the food they wanted to eat. The old man bought knives and axes for his work in the marsh. The old woman bought new plates, glasses, pots, and pans. Life was good. Life was easy.

Eventually, though, the money ran out. And so once again the girl shut herself in her room. When the old couple woke, they could hear the *clickety clack, clickety clack* of the loom. They heard it as they ate breakfast. They heard it as they ate lunch. They heard it as they ate dinner. They heard it late into the night. As the day progressed, the old man and old woman grew increasingly curious. They had so many questions. What did their daughter use for thread? How did she string her loom? Where did she get supplies? She never purchased any in the marketplace. How did she work all day without stopping? It was a mystery.

"Wife," said the old man, "I must know how our daughter creates such unusual and glorious weavings."

"But Husband, we promised never to interrupt her while she works," reminded the old woman.

"I know," said the old man, "but there is a small crack in her wall. We could peek through it. One quick look—she will never notice us."

"Perhaps you are right," said the old woman. "You know, she has never worked this long before."

Quietly, the old man and the old woman crept up to the crack in the wall and looked into their daughter's room. What did they see?

At the loom sat a large white crane. Using her long beak, the crane was plucking feathers from her chest and wings. Then taking her feathers and using them as threads she began weaving. As she fed each pure white feather into the loom, it burst into color. The old couple could not stop staring at this fantastic sight. Suddenly, the crane turned her head and one huge black eye gazed at the crack. The old couple jumped back from the wall, but it was too late. The crane knew what they had done.

With fear in their hearts, the old couple waited for their daughter to leave her room. After a time, the door opened and out walked the girl. She looked into her parents' frightened eyes.

"Father," she said, "dear Father, I am the crane you rescued in the marsh. I wanted—indeed, I needed—to thank you for your great kindness. That is why I came to live with you. I would have been happy forever sharing my life with you and your dear wife, but now it cannot be. You have seen my true form and so I must leave and never return."

"Daughter, do not abandon us," begged the old man.

"We will never reveal your secret," said the old woman. "We love you. We cannot imagine life without you. Please, stay with us."

The girl shook her head. Then, without saying a word, she handed her parents her last weaving. It was dazzling. It showed the marsh and the old couple's cottage at sunset. A flock of cranes was flying through the orange-and-purple sky. Each bird had its own look, its own expression. And there, standing at the cottage door and gazing up at the sky, were the old man and the old woman. The woman had a tear in her eye—a tear as bright as a diamond. Looking at the cloth the old couple could almost smell the trees and feel the late summer breeze.

After handing this weaving to her parents, the girl walked out of the house. As the old couple watched, she gracefully raised her arms and stretched them far above her head. And then, she changed. She was no longer a girl. She was a large beautiful crane. Turning to give her parents one last look, the magnificent bird rose into the air and flew away into the predawn sky.

Each year, when cranes travel from the north to the south, the old man and the old woman stand at their cottage door. And each year, one crane in the flock—a large white bird—flies once, twice, and then three times around the cottage, while the old man and the old woman weep tears even brighter than diamonds.

Notes

About the Story

The Crane proves that not all tales end "happily ever after." There are, of course, many stories of humans who must spend time as animals. Here is a story of an animal who becomes human. It is an unusual twist on a common fairy-tale theme.

For my retelling of this tale, I referred to one version of the story.

> Martin, Rafe. *Mysterious Tales of Japan*. New York: G.P. Putnam's Sons, 1996, 41-46.

If your students are unfamiliar with what a crane looks like, you might want to show them a picture of this lanky bird before sharing the story.

Your Thoughts

The Crane
Writing Names

After determining the number of letters they can write in 30 seconds, students use their data to predict how many letters they can write in one-, two-, and three-minute periods. They can double their initial data to make a prediction for one minute of writing. Then they can double and triple their one-minute predictions to make predictions for two and three minutes respectively. You can discover a great deal about students' mathematical thinking by having them verbalize their methods for making predictions. Each pair of students will need a stopwatch or a watch with a second hand for this activity.

The day her daughter left, the old woman found one snow-white feather next to the loom. The woman decided to use this feather as a pen. With it she wrote her daughter's name over and over again. She hoped that writing would make her feel better, but it didn't. Then her husband thought of a way to cheer up his wife. He turned the name-writing into a game.

He had his wife write their daughter's name over and over for 30 seconds. Next he counted how many letters she managed to compose in that time. Together the old couple predicted how many letters the woman would form if she wrote the name for one minute, then for two minutes, and then for three minutes.

After making these predictions, the old woman started writing—first for one minute, then two minutes, then three. After the woman wrote for one minute, the old couple compared their prediction with the actual results. They did the same after the old woman wrote for two minutes and again after she wrote for three minutes.

You can play this game, too. Have a partner keep track of the time while you write your name over and over for 30 seconds. Next, count how many letters you wrote. With your partner, predict how many letters you think you can write in one minute. After predicting, write and then count the number of letters you formed. Did your prediction come close to the number of letters you actually wrote?

Now predict how many letters you can write in two minutes, then in three minutes. Finally, test each prediction.

When your turn is over, trade roles. Let your partner write his or her name while you keep track of the time.

Weaving

Students will get some artistic experience expressing fraction concepts as they color weaving designs on the next two pages.

After completing their first drawing, students count to determine how many of the 64 squares they colored with each crayon. This helps them understand that each amount represents a fractional part of the whole.

Before beginning the second drawing, students must determine how many squares are equivalent to *one half, one fourth*, and *one eighth* of 64. To do this, students can divide the grid into two equal parts. Counting the squares, they will discover that there are 32 squares in one half. They can divide one half in half again and count to find 16 squares in one fourth. Then they can divide one fourth in half to find 8 squares in one eighth.

You can divide up the grid with your class by projecting a transparency of the grid on an overhead projector. With your guidance the divisions should go smoothly. Once your students know that there are 32 squares in one half of the grid, make sure they understand that they may color any 32 squares with their first color, and, of course, any 16 squares with their second color, and so on.

Try making pictures for weavings of your own. Here are two blank grids, each composed of 64 squares. Your job is to fill in the grids with color. You can create beautiful designs or patterns—ones you would like to weave, if you could.

On the first grid, you will use six colors. When you complete your work, tell about each of your colors by filling in the blanks in this sentence:

I used the color _____ for _____ of my 64 squares.

On the second grid, make a different picture. This time, use four colors. How many squares will be in one half of your picture? Use the first color for that many squares. How many squares will be in one fourth of your picture? Use the second color for that many squares. How many squares will be in one eighth of your picture? Use the third color for that many squares. Then color the same number of squares again, this time using the fourth color.

Which of your two designs is your favorite? Why?

Weaving – 1

Use 6 colors to make a design for a weaving.

<table>
<tr><td></td><td></td><td></td><td></td><td></td><td></td><td></td><td></td></tr>
<tr><td></td><td></td><td></td><td></td><td></td><td></td><td></td><td></td></tr>
<tr><td></td><td></td><td></td><td></td><td></td><td></td><td></td><td></td></tr>
<tr><td></td><td></td><td></td><td></td><td></td><td></td><td></td><td></td></tr>
<tr><td></td><td></td><td></td><td></td><td></td><td></td><td></td><td></td></tr>
<tr><td></td><td></td><td></td><td></td><td></td><td></td><td></td><td></td></tr>
<tr><td></td><td></td><td></td><td></td><td></td><td></td><td></td><td></td></tr>
<tr><td></td><td></td><td></td><td></td><td></td><td></td><td></td><td></td></tr>
</table>

I used the color _____ for _____ of my 64 squares.

I used the color _____ for _____ of my 64 squares.

I used the color _____ for _____ of my 64 squares.

I used the color _____ for _____ of my 64 squares.

I used the color _____ for _____ of my 64 squares.

I used the color _____ for _____ of my 64 squares.

Weaving – 2

Use 4 colors to make a design for a weaving.

Color ½ of the squares in the first color.
Color ¼ of the squares in the second color.
Color ⅛ of the squares in the third color.
Color another ⅛ of the squares in the fourth color.

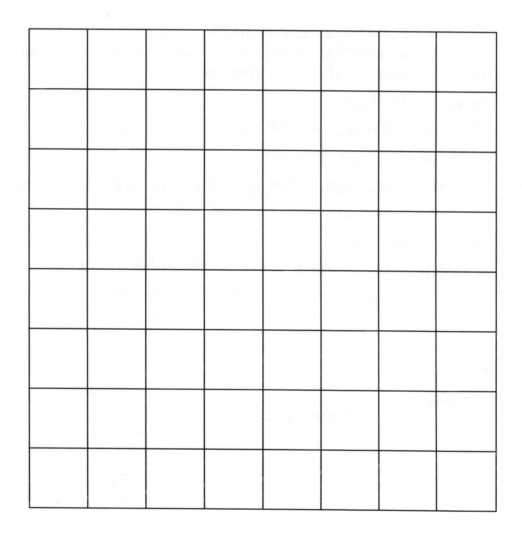

Silver Coins

This is a lesson in logical thinking. By analyzing and organizing numerical information, students discover that the crane dropped 237 coins. After your students solve the problem, let them explain (either through discussion or in writing) how they arrived at their solutions. Then encourage them to make up their own "dropped-coin" problems.

One year the crane dropped a bagful of silver coins outside the old couple's cottage. Follow these clues and you can figure out how many coins the crane dropped.

Here are the clues:

1. There were more than 100 coins.

2. There were fewer than 400 coins.

3. Add all three digits in the number and you get a sum of 12.

4. The hundreds digit is an even number.

5. The sum of the hundreds and the tens digits is 5.

6. The tens digit is one greater than the hundreds digit.

7. The ones digit is four greater than the tens digit.

The Crane

Silver Coins

One year the crane dropped a bagful of silver coins outside the old couple's cottage.

Figure out how many coins the crane dropped.

Here are your clues:

1. There were more than 100 coins.

2. There were fewer than 400 coins.

3. Add all three digits and you get a sum of 12.

4. The hundreds digit is even.

5. The sum of the hundreds and tens digits is 5.

6. The tens digit is one greater than the hundreds digit.

7. The ones digit is four greater than the tens digit.

The Crane

Times for Weaving

This is a two-part problem. In the first part, students must figure out that the old woman wove a total of 20 hours during the first ten days of September, a total of 30 hours during the second ten days, and a total of 40 hours during the third ten days. Students will probably use trial-and-error coupled with some mathematical reasoning to solve this part of the problem.

In the second part, students must create a daily weaving schedule that reflects the number of hours the old woman weaves during each 10-day period. There are many correct ways in which this can be done. Arriving at any of them involves a considerable amount of addition and/or multiplication.

The old woman missed her daughter very much. To comfort herself, the woman started weaving every day.

In September, she wove for a total of 90 hours. She started by weaving a certain number of hours over the first ten days of the month. During the second ten days, she wove that amount plus an additional ten hours. During the third ten days, she wove as much as she had during the second ten days plus ten hours *more*! How many hours did she weave from September 1st to September 10th? How many hours did she weave from the 11th to the 20th? How many hours did she weave from the 21st to the 30th?

She wove at least one hour each day. She never wove more than five hours a day. She never wove the same number of hours for two days in a row.

Create a weaving schedule for the old woman. Show how many hours you think she worked on her weaving each day in September.

Times for Weaving

Show the old woman's September weaving schedule by filling in this calendar page.

September

Sunday	Monday	Tuesday	Wednesday	Thursday	Friday	Saturday
1 ___hours	2 ___hours	3 ___hours	4 ___hours	5 ___hours	6 ___hours	7 ___hours
8 ___hours	9 ___hours	10 ___hours	11 ___hours	12 ___hours	13 ___hours	14 ___hours
15 ___hours	16 ___hours	17 ___hours	18 ___hours	19 ___hours	20 ___hours	21 ___hours
22 ___hours	23 ___hours	24 ___hours	25 ___hours	26 ___hours	27 ___hours	28 ___hours
29 ___hours	30 ___hours					

Wiley and the Hairy Man

A Tale from the United States

Once upon a long time ago, Wiley, his mama, and their hound dogs lived in a forest near a dark, dark swamp. No other sensible creatures dwelt anywhere nearby. Why? Hairy Man. Yes, Hairy Man himself lived in that swampland. Knowledgeable people stay away from Hairy Man 'cause if he gets you, you are gone for sure. So how come Wiley and his Mama made their home so near that scary place? Well, Wiley's Mama, she knew some conjuring—some good magic. She learned from her mama, an old woman who lived on the Tombigbee River. Bit by bit, Wiley's mama taught her son all he needed to know.

She made certain, for instance, that Wiley always left home with his dogs—knowing as she did that Hairy Man hates, loathes, cannot tolerate hound dogs.

One muggy morning, though, things did not go as planned.

"Wiley," his mama said after breakfast, "take your hounds into the forest and cut us some bamboo poles. We need to build a new hen house."

So Wiley grabbed his ax, called his dogs, and set out for work. He walked here and there until he found a good cluster of bamboo. Then he started chopping. He hadn't been toiling long when a wild pig came charging through the bushes. What did those hound dogs do when they saw that pig running? They chased it, of course. Gone. Those dogs were gone so far, Wiley couldn't even hear them bark. Wiley was not happy

about this. No, he was not. But before he could call the hounds back to his side, Hairy Man came tramping through the trees.

Hairy Man was furry from head to hoof. Yes, he had hooves like a horse or a cow. He also had blood-red eyes and pointy yellow teeth. Imagine the ugliest creature you can and then make the creature even uglier. That's Hairy Man.

"Hairy Man," Wiley cried out, "keep away from me!"

"Why should I?" asked Hairy Man. "Your hounds are gone. Chasing a pig, if I'm not mistaken. But Wiley, let's not waste time. I have my capturing bag here and I plan to put you in it."

When Wiley heard these words, he dropped his ax and climbed to the top of a tall, tall tree.

"With those hooves, Hairy Man, you can't get up here," called Wiley.

"Agreed," said Hairy Man. "But with your ax, I can hack you down to the ground."

Hairy Man picked up the ax and started chopping and wood chips started flying. But every time he chopped, Wiley called out, "Fly chips. Fly back to your same old place."

Words said is magic done, and so the chips flew back into the tree trunk. Hairy Man kept chopping and Wiley kept ordering wood chips back into place. On and on it went until Wiley heard his dogs yelping in the distance. Then he called out, "Dogs, come here to me!"

Immediately, the dogs came barking and

snarling. When Hairy Man heard those sounds, he stamped and stomped his way back to the swamp.

Wiley went home and told his mama everything that had happened.

"You did good, Wiley," said Mama. "You tricked Hairy Man once. But we've got to trick him three times, three times, Wiley, and then Hairy Man won't trouble us anymore."

"But how can I trick him again, Mama?" asked Wiley.

After pondering on this a bit, Mama told Wiley just what to do.

The next morning, Wiley tied his hounds with a rope so they could not leave their pen. Then he went down to the rapid river that runs through the forest. Sure enough, along came Hairy Man—happy as can be to see Wiley without his snarling dogs.

"Hairy Man," said Wiley, "I understand you can do some powerful conjuring."

Hairy Man smiled and said, "Wiley, I happen to be the best conjurer ever. I can do anything I like with conjuring."

"Can you turn yourself into a giraffe?" asked Wiley.

"Nothing easier," said Hairy Man and he spun himself into that long-necked beast.

"Very impressive," said Wiley, "but I doubt you can turn into an alligator."

"Of course I can," said Hairy Man as he stretched into a 'gator's knobby form.

"How about an elephant?" asked Wiley.

"So be it," said Hairy Man. And so it was.

"Sure, you can turn into big animals. But I wonder, can you turn into a small critter, a possum, for instance?" asked Wiley.

"Just watch," said Hairy Man.

Wiley did watch. The moment Hairy Man

become a possum, Wiley seized the tiny animal and threw it into the capturing bag. He tied the sack up tight and tossed it into the river.

This done, Wiley started home, whistling a joyous tune. He stopped his song though, when suddenly, he saw Hairy Man directly in front of him.

"Hairy Man, how did you get out of that bag?" asked Wiley.

"I changed into the wind and blew free," smirked Hairy Man.

Wiley thought quick and then he said, "Hairy Man, you showed me some fine conjuring, but it was all here-and-now magic. I doubt you can do far-and-wide magic."

"Try me," said Hairy Man.

"I expect you can make the rope belt on my pants disappear, it being here and now," said Wiley.

"You are right about that," said Hairy Man.

"But no way can you make all the rope in the county disappear, 'cause that's far and wide," said Wiley.

"Ha," said Hairy Man, "I can, and I will!"

Instantly Wiley's rope belt vanished and his pants fell down around his knees. Hairy Man laughed and laughed, but not for long. You see, at that exact moment the rope tying Wiley's hound dogs also disappeared! Those dogs came growling straight for Hairy Man. One eye-blink later and Hairy Man was headed for the swamp.

When Mama heard Wiley's tale, she said, "We tricked him twice, Wiley, but to be forever safe, we've got to trick him once more."

Wiley's mama knew that soon—very

soon—Hairy Man would come to their home, right to their door, looking for Wiley. How could she protect her son? She thought and thought until she had a plan.

"Wiley," she said, "lock up our hounds so they can't leave their pen. Then go to the sty, take that baby piglet away from its mama, and bring it here to me."

When Wiley returned with the piglet, Mama placed the little animal in Wiley's bed and pulled blankets all around it. Next she told Wiley to hide behind the cookstove.

It wasn't more than a half hour before Hairy Man came knocking.

"Mama," he called out, "I'm here for your baby. Don't fight me now. Just give that youngster to me."

"Hairy Man," said Mama, "you know I won't give you my baby. So stop asking."

"I will set this house on fire if you don't give me your little one," threatened Hairy Man.

"I've got plenty of milk in here. I'll use it to put out your fire," declared Mama.

"I'll dry up your well," announced Hairy Man.

"I'll dig a new one," answered Mama.

"I'll kill your cows and destroy your crops with boll weevils unless I get your baby," said Hairy Man.

"Oh, please, Hairy Man, don't do that. It is too wicked and too spiteful," cried Mama.

"Wicked and spiteful, that's me," laughed Hairy Man.

"If I give you my baby, will you go away and never come back?" asked Mama.

"Never come back," sneered Hairy Man.

"You win," whimpered Mama. "Take my baby, my poor, poor baby."

Chuckling, Hairy Man walked into the house and over to the bed. He reached down and grabbed the pig. Roused from sleep, the little animal squealed so loudly the room rattled. Glaring down at his prize, Hairy Man shouted, "Mama, this isn't your baby!"

"Why sure it is, Hairy Man. That's my baby pig. And now the pitiful thing is all yours," said Mama.

Hairy Man screamed, yelled, roared. It did no good. For he had been tricked three times, and now Wiley and his mama were three times safe. Defeated, Hairy Man slunk back to his swamp and never troubled Wiley or his Mama again.

Notes

About the Story

Wiley and the Hairy Man is an African-American folk tale. You'll notice that in some ways it resembles *The Legend of Knockmany*. In both stories a clever woman outwits a dangerous foe. In both, a counterfeit baby plays a major role.

Wiley's adventure also has something in common with the classic French story, *Puss in Boots*. In this tale, Puss appeals to an ogre's pride and gets the beast to turn himself into various animals. When the ogre transforms himself into a mouse, Puss eats him.

For this retelling I relied on three sources.

Bang, Molly Garrett. *Wiley and the Hairy Man*. New York: Aladdin Paperbacks Ready-to-Read, 1976.

Botkin, B.A. ed. *A Treasury of American Folklore: The Stories, Legends, Tall Tales, Traditions, Ballads and Songs of the American People*. New York: Bonanza Books, 1983, 682–687 (Originally published in 1944.)

Hamilton, Virginia. *The People Could Fly: American Black Folktales*. New York: Knopf, 1985, 90–103.

Molly Garrett Bang's version is easy enough for most third graders to read independently.

Your Thoughts

Name the Hounds
Feed the Hounds

Here are two problems in deductive reasoning. For each, students must follow a set of clues and record their responses to them on a grid. In one problem, students match each dog with its color. In the other, they match each dog to its preferred food.

If your students have never tackled a logic grid before, they may need help interpreting clues and filling in the grid. You might want to make an overhead transparency of the first grid so you can work with the entire class at once. Students should then be able to complete the second grid independently.

The dog's colors are

Tug: white	Hug: tan	Pug: black	Chug: brown	Dug: red

The foods the dogs eat are

Tug: eggs	Hug: cheese	Pug: chicken	Chug: bones	Dug: liver

Wiley has five hound dogs. Their names Tug, Hug, Pug, Chug, and Dug. Each dog is a different color—black, brown, white, red, and tan.

Your job is to match each dog to its proper color. There are three clues to help you. Read the clues very carefully, and use the information to fill in the logic grid.

As soon as you know that a dog cannot be a certain color, put an X in the box where that name column and color row meet.

As soon as you know that a dog must be a particular color, put a check in the box where that name column and that color row meet. Since no other dog can have that color, fill in the rest of that row with Xs. Since a dog can have only one color, fill in the rest of that column with Xs.

If you think about each clue and carefully fill in the grid, you will discover each dog's color. After you have done this, find out which food each dog likes best.

Name the Hounds

Wiley has five hound dogs. Their names are Tug, Hug, Pug, Chug, and Dug. Each dog is a different color. Their colors are black, brown, white, red, and tan. Use the clues and the grid to match each dog with its proper color.

Here are the clues.

1. Pug is a black dog. She is bigger than both Tug and the tan dog.

2. Pug, Tug, and Hug are faster than both the red dog and the brown dog.

3. Chug can bark louder than the red dog.

	Tug	Hug	Pug	Chug	Dug
Black					
Brown					
White					
Red					
Tan					

Wiley and the Hairy Man

Feed the Hounds

Wiley has five hound dogs. Their names are Tug, Hug, Pug, Chug, and Dug. Each dog has a favorite food—either chicken, eggs, liver, cheese, or bones. Use the clues and the grid to match each dog with its favorite food.

Here are the clues.

1. Tug does not like meat or bones. Neither does Hug.

2. Chug and Pug can both out-race the dog who likes liver.

3. Hug never ever eats eggs.

4. Chug never eats chicken.

	Tug	Hug	Pug	Chug	Dug
Chicken					
Eggs					
Liver					
Cheese					
Bones					

Wiley's Day

Students design a 24-hour schedule of Wiley's activities. They total the amount of time he spends on each activity and then record this information in a 24-hour circle graph. Upon completing their work, students will see how easy it is to interpret information presented in graph form.

After graphing Wiley's day, students might like to graph their own days. Suggest that they make 24-hour graphs entitled "My Most Perfect Day" or "The Most Horrible 24 Hours Imaginable."

Wiley has a very busy day. He does chores. He goes hunting. He plays with his hounds. He eats and sleeps. And he studies conjuring. His mama teaches him some new magic every day and every day he practices and practices.

Design a day for Wiley. You can have him do anything you like. Once you decide what he does during each hour of this day, list his activities, hour by hour, on the first worksheet to complete Wiley's Daily Schedule.

At the top of the second worksheet record the total amount of time Wiley spends at each activity you named. Let's say Wiley spends a total of three hours eating. In that case, write the number of hours, "3," and the activity, "eating," so that one sentence reads, "Wiley spends 3 hours eating."

Now graph this information at the bottom of this worksheet. If Wiley spends 3 hours eating, for example, you would shade three adjacent sections of the graph with a single color—orange, for instance—and label them "EATING." For each of Wiley's activities, shade a matching number of adjacent sections. Use a different color for each activity. Continue in this way until your circle graph is complete.

Wiley's Day – 1

Wiley is very busy every day.

Plan a day for Wiley. He can do whatever you like.

Make a schedule for Wiley. List everything he does during the day.

Wiley's Daily Schedule

8 AM _____	1 AM _____
9 AM _____	2 AM _____
10 AM _____	3 AM _____
11 AM _____	4 AM _____
12 NOON _____	5 AM _____
1 PM _____	6 AM _____
2 PM _____	7 AM _____
3 PM _____	
4 PM _____	
5 PM _____	
6 PM _____	
7 PM _____	
8 PM _____	
9 PM _____	
10 PM _____	
11 PM _____	
12 MIDNIGHT _____	

Wiley and the Hairy Man

Wiley's Day – 2

Write the total number of hours Wiley spends at each of his activities.

Wiley spends

_____ hours _____.   _____ hours _____.

_____ hours _____.   _____ hours _____.

_____ hours _____.   _____ hours _____.

_____ hours _____.   _____ hours _____.

_____ hours _____.   _____ hours _____.

_____ hours _____.   _____ hours _____.

Make a 24-hour circle graph to show Wiley's day.

Use a different color for each activity.

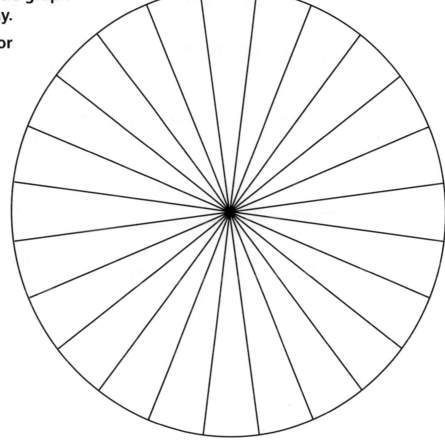

Pumpkin Pie

To solve this problem students must think of a variety of ways to make $5.87 with a specific set of one-dollar bills and coins. There are many possible solutions to this problem. Encourage your students to find as many as they can. Compile a list of all dollar-bill-and-coin combinations your students discover.

There is only one way to pay using the fewest coins: 4 one-dollar bills, 7 quarters, 1 dime, and 2 pennies.

There is only one way to pay using the most coins: 1 one-dollar bill, 9 quarters, 14 dimes, 19 nickels, and 27 pennies.

Mama is so happy that Wiley tricked Hairy Man, she decides to celebrate by making a great big pumpkin pie. She has a great big pumpkin and all the spices, but she does not have enough flour or sugar. She knows that a bag of flour costs $2.36 and that a bag of sugar costs $3.51.

Mama keeps her money in a cookie jar. Emptying the jar, she counts four one-dollar bills, 10 quarters, 15 dimes, 20 nickels, and 30 pennies.

Mama thinks of a way to pay for the food with an exact amount of money. She can use three of the one-dollar bills, eight quarters, eight dimes, one nickel, and two pennies.

Wiley thinks of a different way Mama can pay. He says she can use two one-dollar bills, eight quarters, thirteen dimes, ten nickels, and seven pennies.

Think of more ways Mama can pay for the flour and sugar. How can she pay using the fewest coins? How can she pay using the most coins?

Pumpkin Pie

Mama wants to make a big pumpkin pie to celebrate Wiley's victory over Hairy Man. She must buy a bag of flour and a bag of sugar.

The flour costs $2.36.

The sugar costs $3.51.

In Mama's cookie jar there are 4 $1-bills, 10 quarters, 15 dimes, 20 nickels, and 30 pennies.

How can Mama pay for the flour and sugar using the exact amount of money?

How can she pay using the fewest coins?

How can she pay using the most coins?

Angry, Angry Hairy Man

To solve the two parts of this challenging problem, your students must take several steps. First, they must figure out that because there are 6 ten-minute periods in one hour, there are 144 ten-minute periods in 24 hours. Hairy Man, therefore, screams 144 times. The pig squeals for 143 minutes. That's a total of 2 hours, 23 minutes.

These problems can be solved by multiplying, but they can also be solved less formally by adding. Pictures and manipulative materials will prove very helpful as children work their way through Hairy Man's temper tantrum. Some children will ask for, and should be given, calculators.

When Hairy Man got back to the swamp, he was angry. He had never been so angry before. Every ten minutes on the dot his rage built up and he howled louder than a wounded hound dog. When Hairy Man howled, the poor little pig squealed in terror. She squealed for one full minute after every howl.

All this howling and squealing happened every ten minutes for exactly twenty-four hours.

Can you figure out how many times Hairy Man howled and how long the pig squealed during that noisy time?

Angry, Angry Hairy Man

When Hairy Man got back to the swamp he was very angry. He had one way to express his anger and that was to howl out loud.

That angry Hairy Man howled every ten minutes for 24 hours.

After each howl, the pig was so scared she squealed for one minute—without stopping—every ten minutes for 24 hours.

During that noisy 24 hours, how many times did Hairy Man howl?

And how much time did the pig spend squealing?

A Tale from Germany

Once upon a time an old woman and an old man lived on the edge of a wild, wild woods. The couple loved each other and were very happy, but they were not perfectly happy because they did not have a child.

It so happened that the old couple lived next to a witch. This witch had a beautiful garden and in it she grew lettuce—lots of lettuce and only lettuce.

Every day the old woman saw the lettuce shining in the sunlight. She craved one taste, just one taste. Day after day, her craving grew and grew. Soon she became so sad, she lost interest in all other food and refused to eat anything. The old man feared his wife would die unless she had a bit of the witch's greenery. That is why, late one night, the old man tiptoed into the witch's garden and picked a handful of sweet lettuce leaves. The witch saw the old man and stopped him.

"What are you doing in my garden? What are you doing with my plants?" the witch asked.

"Please," said the old man, "let me have a few leaves of your lettuce. Without it, my wife will surely die."

"You want my lettuce?" said the witch. "Very well, but I must have something in return. If your wife ever has a baby, you must give the child to me."

The old man agreed to the witch's demand. After all, he did not believe that his wife would ever have a child. And the dear woman did need this lettuce. Yes, she needed it very much. When the old man got home, his wife ate every leaf of the lettuce and was soon happy again.

Months later, as if by magic, the old woman gave birth to a beautiful baby girl. On the morning the baby was born, the witch knocked on the old couple's door. She had come for the child. The old couple had no choice but to give up their little girl.

The witch named the infant Rapunzel. Odd as it may seem, the witch truly loved the girl and took good care of her. Because she did not want to share her daughter with anyone else in the world, the witch built a tall tower in the middle of the wild, wild woods. At the very top there was a room—the only room in the tower. In this room there was a single window. On Rapunzel's twelfth birthday, the witch took the child to the tower to live.

Every day the witch went to the tower to visit Rapunzel. But how did she reach the tower room? She did not climb stairs. Oh no, there were no stairs. Instead, she would stand beneath the tower window and call, "Rapunzel, Rapunzel, let down your hair."

Hair—Rapunzel had a very long, exceptionally long, extraordinarily long hair which she kept in a braid. When she heard the witch's voice, Rapunzel dropped her braided locks out the window. The braid almost reached the ground. Then the witch, using Rapunzel's hair as a rope, climbed up the tower wall and crawled through the window to join her daughter.

Several years passed. One beautiful spring afternoon, a prince was traveling through the wild, wild woods when he approached the tall tower in the middle of a circle of trees. As he stood wondering about this odd building, the witch appeared. Quickly, the prince hid behind some bushes. From there, he could hear the witch call Rapunzel. He could see the witch climb the tower. Then he waited until the witch came down and walked away through the woods. When she was gone the prince called out, in a witch-like voice, "Rapunzel, Rapunzel, let down your hair."

Moments later, the prince arrived on the windowsill. Imagine Rapunzel's surprise when she saw a handsome young man instead of her witch mother. At first Rapunzel was scared, but the prince was so kind that soon he and Rapunzel became good friends. Several hours later, the prince left the tower, but he promised to return. And he did return, day after day.

In a short time, Rapunzel and the prince fell in love. They wanted to get married and live together in the prince's castle. They had one problem, though. How could they escape the tower together? Needless to say, Rapunzel could not climb down her own braid. Fortunately, the prince thought of a plan. Each time he came to visit, he would bring Rapunzel a skein of wool. Rapunzel would twist the skeins, and eventually they would have a braided rope as long and as strong as Rapunzel's hair. Then they would hang the rope from a hook next to the window and climb down the tower together.

It was a fine plan, but something went terribly wrong. You see, one day, after the witch slid through the window, Rapunzel said,

"It's strange, Mother, how much heavier you are than my prince."

Now the witch was surprised. She made Rapunzel tell her all about the prince. The witch listened and grew angry, very angry. In her rage, she cut off Rapunzel's magnificent braid and attached it to the window hook. Poking and pinching, she forced Rapunzel out the window and down the braid. When they were on the ground, the witch dragged Rapunzel into the deepest part of the woods and left her in a cave.

Then the witch returned to the tower and waited for the prince. When he called, she dropped Rapunzel's braid out the window. The prince started climbing, but before he got to the top, the witch reached out and pushed him hard—very hard. The prince fell to earth. He landed in a thorn bush and the sharp, pointed thorns stabbed his eyes. The prince was blind. He stumbled into the wild, wild woods, calling for Rapunzel.

For more than a year, the prince wandered the woods, blind and alone. One evening, though, he passed very near Rapunzel's cave, and there he heard his true love's voice. She was singing a sad song while picking berries for her supper. The prince shouted her name, and Rapunzel rushed to his side. The blind prince was so thin and ragged, Rapunzel could not help but cry as she held him tight. Her tears fell into the prince's eyes. Something magical happened at that moment—the prince could see again. He saw Rapunzel's beautiful face. He saw the trees. He saw the birds. He saw everything.

Now, together at last, the prince and Rapunzel left the wild, wild woods and walked to the prince's kingdom, his castle, and his family.

The king and queen were overjoyed to see their son. The prince announced that he loved Rapunzel and wanted to marry her. So the king and queen planned a huge wedding and invited everyone in the kingdom to celebrate. From that day on, Rapunzel and the prince lived happily ever after.

Notes

About the Story

Rapunzel is a classic fairy tale. Many students will already be familiar with the story. Even so, they won't object to hearing it again.

Some students may wonder what happened to the witch. The Grimms do not tell us. This means that students can make up their own minds—and the issue might warrant some discussion. Does the witch look for another child? Does she join a family of witches? Does she engage in a slew of evil doings? You may want to encourage students to write stories telling what they think happened.

For this retelling, I relied on three versions of the tale.

Grimm, Jacob and Wilhelm. *The Complete Grimm's Fairy Tales*. New York: Pantheon Books, 1972, 73–76.

Grimm, Jacob and Wilhelm. Lore Segal and Maurice Sendak, eds. Lore Segal and Randall Jarrell, trans. *The Juniper Tree*. New York: Farrar, Straus and Giroux, 1973, 247–255.

Lang, Andrew, ed. *The Rainbow Fairy Book*. New York: William Morrow and Co., Books of Wonder, 1993, 11–17.

Your Thoughts

Rapunzel

Rapunzel's Hair

In the first part of this problem, students discover the length of Rapunzel's hair after one year. Rapunzel's hair grows 50 cm a month—that's 1 m every two months, or 6 m every year. Students add the 6 m to Rapunzel's already 1-m-long hair to get 7 m.

For the second part of the problem, students must do three things—determine the length of Rapunzel's hair after another year passes (13 m), determine one year's hair growth for Joy (3 m), and track both Rapunzel's and Joy's hair growth for four more years. They find that on Joy's fifth birthday her hair measures 15 m and Rapunzel's measures 37 m.

The third part of the problem has students measure a 37-m distance. Provide them with a meter stick and chalk and allow them to measure the distance in a hallway or on the playground.

Rapunzel did not have regular everyday hair. No indeed. The witch wanted Rapunzel's hair to grow fast—very fast. So when Rapunzel was two years old, the witch zapped her with a super-strong hair-growing spell. As a result, Rapunzel's hair grew 50 cm every month.

Rapunzel married the prince on January 1st. That day, her hair was 1 m long. Exactly one year later she gave birth to a beautiful baby girl and named the child Joy. How long was Rapunzel's hair on the day Joy was born?

Rapunzel was amazed when Joy's hair started growing fast—very fast. You see, Joy had inherited the magic spell, but the spell was not as strong. It made Joy's hair grow exactly half as fast as her mother's. On Joy's first birthday, Rapunzel measured her daughter's hair. How long had it grown in one year?

How long was Joy's hair on her fifth birthday? How long was Rapunzel's hair on that same day?

If, on Joy's fifth birthday, Rapunzel came to your classroom and stood in the doorway, how far down the hall would her hair stretch?

Draw your own picture of Rapunzel and Joy.

Rapunzel's Hair

Rapunzel's amazing hair grows 50 cm every month. On January 1st, Rapunzel married the prince. That day her hair was 1 m long. Exactly one year later, Rapunzel gave birth to a beautiful baby girl.

How long was Rapunzel's hair on the day her baby was born?

Rapunzel named her baby Joy. Joy's hair grows fast, too, but only half as fast as Rapunzel's.

How long was Joy's hair on her first birthday? How long was it on her fifth birthday? How long was Rapunzel's hair on that day?

If, on Joy's fifth birthday, Rapunzel came to your classroom and stood in the doorway, how far down the hall would her hair stretch?

Draw your own picture of Rapunzel and Joy.

Rapunzel

Back to the Tower

This problem reinforces students' understanding of multiples of a number and common multiples. Students must record four sets of numbers and observe where the sets intersect. To do this, they may choose to model the 105 steps on a slightly extended hundreds board. Alternatively, they may make four lists of numbers, one for each kind of decoration, and then determine which numbers appear on more than one list.

Stars and flowers appear on steps 6, 12, 18, 24, 30, 36, 42, 48, 54, 60, 66, 72, 78, 84, 90, 96, and 102. Flowers and crescents appear on steps 15, 30, 45, 60, 75, 90, and 105. Stars, flowers, and diamonds appear on steps 42 and 84. Crescents, flowers, and diamonds appear on step 105. None of the steps have all four kinds of decorations.

Soon after they were married, the prince and Rapunzel realized they missed spending time in their old meeting place, the tower. So they created a path from the castle through the wild, wild woods back to Rapunzel's old home. Now they could get there easily, but how could they get up to the window?

They built a stairway—and a long stairway it was. Indeed, it had 105 steps. Rapunzel liked the new steps so much, she decided to decorate them.

Counting upward from the bottom step, Rapunzel painted stars on every second step. She painted flowers on every third step, crescent moons on every fifth step, and diamonds on every seventh step.

How many steps had stars and flowers? How many steps had flowers and crescents? How many steps had stars, flowers, and diamonds? How many steps had crescents, flowers, and diamonds? How many had all four decorations?

Rapunzel

Back to the Tower

The prince and Rapunzel missed the tower, so they made a path from the castle through the wild, wild woods back to it. They built a long stairway up to the window. Rapunzel decided to decorate her new stairway. The stairway had 105 steps.

Counting from the bottom step:

Rapunzel painted stars on every other step.

She painted flowers on every third step.

She painted crescent moons on every fifth step.

She painted diamonds on every seventh step.

- How many steps had stars and flowers?
- How many steps had flowers and crescents?
- How many steps had stars, flowers, and diamonds?
- How many steps had crescents, flowers, and diamonds?
- How many had all four decorations?

Rapunzel

Lettuce Recipes

There are two parts to this problem. First, students must first use logical reasoning to arrange the seven kinds of recipes in order from least to greatest—cakes, casseroles, soups, breads, pies, salads, and cookies. Students must then partition the 324 recipes into seven groups, each having from 41 to 89 items. In comparing their work, students will realize that a great many possible recipe combinations can satisfy the cook's desires.

Rapunzel loved lettuce. She really loved lettuce. She loved it so much that she wanted some at every meal. She ate lettuce in salads, soups, and casseroles. She gobbled lettuce cakes, lettuce pies, lettuce breads, and lettuce cookies.

To delight Rapunzel, the royal cook invented new lettuce recipes every day. Eventually the cook had so many recipes that she decided to write a lettuce cookbook. The cook wanted the book to have 324 recipes. She wanted recipes for salads, soups, casseroles, cakes, pies, breads, and cookies. That's seven kinds of lettuce recipes.

The cook had very strict rules for her book. She wanted there to be fewer cake recipes than any other kind of recipe, more cookie recipes than any other kind, fewer soup than bread recipes, more pie than bread recipes, one more soup recipe than casserole recipes, more salad than pie recipes, more than 40 cake recipes, and fewer than 90 cookie recipes.

Following these rules, how many of each kind of recipe could the cook include in her cookbook? Compare your answer with those of your classmates. How are the answers alike? How are they different? How do you explain this?

Rapunzel

Lettuce Recipes

The royal cook loved inventing lettuce recipes for Rapunzel. Finally, the cook had so many recipes she decided to write a cookbook.

The cook wanted her book to have 324 lettuce recipes. She wanted seven kinds of recipes—recipes for salads, soups, casseroles, cakes, pies, breads, and cookies.

- She wanted fewer cake recipes than any other kind.

- She wanted more cookie recipes than any other kind.

- She wanted fewer soup than bread recipes.

- She wanted more pie than bread recipes.

- She wanted one more soup than casserole recipes.

- She wanted more salad than pie recipes.

- She wanted more than 40 cake recipes.

- She wanted fewer than 90 cookie recipes.

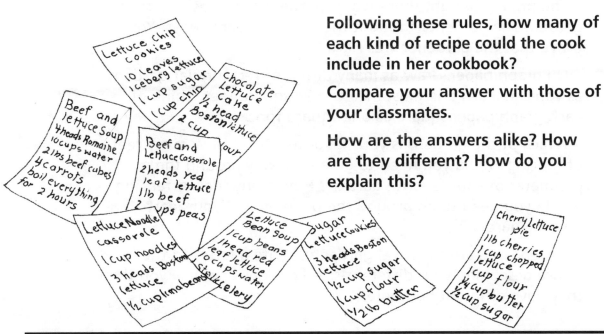

Following these rules, how many of each kind of recipe could the cook include in her cookbook?

Compare your answer with those of your classmates.

How are the answers alike? How are they different? How do you explain this?

Rapunzel

The Lettuce Garden

Challenged by having to find different-sized rectangular garden plots, all with the same area—24 square feet—students outline the gardens, creating these arrays: 1 x 24, 2 x 12, 3 x 8, and 4 x 6. Consider having students first model the gardens using 1-inch-square Color Tiles or other square manipulatives. Then they can draw the gardens on the graph paper provided.

Students may be surprised to discover that although the area is consistent, the perimeter of the gardens varies. So, the 1 x 24 garden has a perimeter of 50 ft; the 2 x 12 has a perimeter of 28 ft, the 3 x 8 has a perimeter of 22 ft, and the 4 x 6 has a perimeter of 20 ft.

Writing the letter to the prince will help students verbalize their mathematical insights.

Shortly after their wedding, the prince discovered that Rapunzel had a very strange eating habit. She loved one food more than all others. Lettuce. Yes, Rapunzel ate lettuce for breakfast, lunch, and dinner. The prince wanted to satisfy Rapunzel's every desire. That's why he ordered the royal gardener to create a 24-square-foot rectangular garden solely devoted to growing lettuce.

The prince thought there was just one way to design such a garden. The gardener thought there were several ways. You can prove that the gardener was right.

On graph paper, draw as many different rectangular gardens as you can. Each one must measure 24 square feet. Think of each graph paper square as one square foot.

Which of the gardens is your favorite? Why? How many feet of fencing would you need to go all the way around the perimeter of your favorite garden? How many feet of fencing would you need to go around the perimeter of your least favorite garden?

Pretend you are the royal gardener. Write a letter to the prince describing your different-sized designs and why you think they have different perimeters.

Rapunzel

The Lettuce Garden – 1

The prince ordered the royal gardener to create a lettuce garden. The garden had to be rectangular and it had to measure 24 square feet.

The prince thought there was only one way to make such a garden. The gardener thought there were more ways. Prove that the gardener was right.

Design as many different-sized gardens as you can on graph paper. Make sure each garden measures 24 square feet.

Which garden is your favorite? Why?

How many feet of fencing would you need to go around the perimeter of your favorite garden? How many feet would you need for the perimeter of your least favorite garden?

Pretend you are the royal gardener. Write a letter to the prince describing your different-sized gardens and why you think they have different perimeters.

Rapunzel

The Lettuce Garden – 2

Design your gardens on this paper. Think of each small square as 1 square foot.

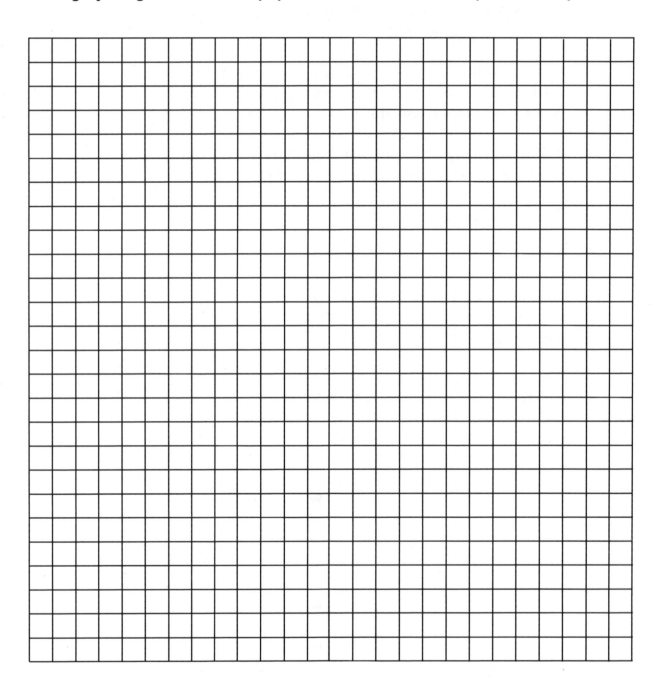

A Legend of Knockmany

A Tale from Ireland

Fin M'Coul was a very strong giant. He could yank towering oak trees from the ground, juggle four huge boulders in the air, and carry seven men on his back. Yes, Fin was powerfully strong and no other giants dared fight him.

Well, actually, that's not completely true. There was one giant, Cucullin by name, who was bigger and stronger than Fin. When Cucullin stamped his foot, the land shook for miles around. On stormy days, he grabbed lightning bolts from the sky and used them to scratch his back. But to prove he was the most fearsome of giants, Cucullin had to beat Fin in a fight.

Fin, strong as he was, feared battling Cucullin. So he built a home on top of Knockmany Mountain. From his front yard he could see far and wide, wide and far. If Cucullin approached within 100 miles of Knockmany, Fin would know.

Now Fin had a wife named Oonagh and a fine giantess she was. She could tell when Fin had a worry and, somehow, she managed to solve all his problems. That's why, one sunny morning, Oonagh insisted on knowing what dark, gloomy thoughts were disturbing the mood of her darling husband.

"Tell me, my sweet, what could be troubling you this lovely day?"

"Oonagh, my Oonagh, the time has come. Cucullin, that monster of a giant, is heading toward Knockmany Mountain. Bad luck—sad luck—either I fight him, and most probably lose my life in the process, or I leave our home and hide away for the rest of my days."

Oonagh was not flustered by this news. All she said was, "Suck on your thumb, my dear, and tell me when Cucullin will be at our door."

Does that sound like an odd request? I'm sure it does. But you see, Fin had a magic thumb and when he sucked it, he could glimpse the future.

So Fin sucked and then announced, "Two o'clock tomorrow afternoon—that's when Cucullin arrives and I better be long gone by then."

"Stay just where you are," said Oonagh. "I'll rid us of that brute Cucullin. Leave everything to me."

Fin knew his wife was clever. Maybe, just maybe, she was quick-witted enough to defeat his enemy.

After Fin settled in his favorite chair, Oonagh went to her weaving shed. There she selected nine threads of wool, each in a different color. She braided them into three separate plaits with three colors in each one. Then she tied one plait around her right arm, one across her chest directly over her heart, and one around her right ankle. She knew that with these colorful threads in place, she would have bright thoughts and would succeed in any endeavor. Indeed, by the time she finished braiding, she had a plan—a plan to defeat Cucullin.

Early the next morning, Oonagh called on

her neighbors, and from each she borrowed an iron griddle. When she had 21 griddles under her arm, she returned home and started making bread dough—lots and lots of bread dough—enough for 22 loaves. She hid an iron griddle in the center of each of 21 loaves and then, after heating her oven, she baked those 21 iron-filled loaves along with the one ordinary loaf.

Next she made a bowl of curds and whey and showed Fin how they would help in her plan. Finally she dragged an old cradle into the living room and filled it with bright blue blankets. As two o'clock approached, Fin and Oonagh looked out their window and saw Cucullin crossing the valley.

"He will be here soon," said Oonagh. "So Fin, my dear, do as I say and without a complaint. Hop into the cradle and let me wrap you in covers. I shall entertain Cucullin while you pretend to be our own bit of a baby boy. Pay close attention to all I do and say. When the time is right, you will know how to act."

Grumbling a bit, Fin did as his wife ordered. He climbed into the cradle and lay there sucking his magic thumb hoping to discover what the future would bring.

Soon enough Cucullin pounded on the front door. "I've come for Fin M'Coul," the hulking giant said when Oonagh answered.

"Well, my good fellow, I'm Mrs. M'Coul, but you won't find my husband at home. He ran off in a fit of anger. It seems some fool of a giant named Cucullin is looking to fight him. If my Fin ever catches this Cucullin, I feel sorry for his poor wife having to nurse her man back to health," said Oonagh as she stepped outside her house.

"It happens I'm Cucullin, and I do want to fight Fin. As for nursing, I'll see to it that your husband gets his time in a sick bed," bragged Cucullin.

"I don't suppose you have ever seen my Fin or you would not be so eager to fight him. But if you insist on waiting, do you think you could do me one small favor?" asked Oonagh.

"Certainly," said Cucullin.

"A cold west wind is coming through the door. If Fin were about, he would pick up the house and twist it around so that the breeze would not give our baby a chill. I'd like you to manage this little chore for me, if you don't mind," said Oonagh.

"Hmmmm," thought Cucullin, "Fin can lift his house. Well, if he can, I can, too."

As it happens, Cucullin had his own magic finger. His finger did not tell the future, though. Instead, it held all of Cucullin's vast strength. The more he pulled on it, the stronger he got.

To meet Oonagh's request, Cucullin tugged his magic finger three times. Then he lifted the house into the air, spun it about, and placed it back on the ground.

"Much better," said Oonagh. Then she continued, "Since we happen to be outside, I wonder, could you do me one more favor? I'm so tired of walking down the mountain to get water. This morning Fin promised to dig me a well right here next to our house. He said it would only take a few minutes to tunnel through the rock, but off he ran before doing the job. Perhaps, while you're waiting, you could take care of this task."

"Hmmmm," thought Cucullin, "Fin digs through rocks. Well, if he can, I can, too."

So Cucullin pulled on his magic finger nine

times and then started digging. He plowed through 400 feet of mountain before hitting water.

After admiring her new well, Oonagh led Cucullin into the kitchen and said, "Let me thank you properly for your help. I have warm bread fresh from the oven. It's yours to enjoy."

Hungry as could be after all his work, Cucullin grabbed a loaf, shoved the whole thing in his mouth, and chewed.

"Blood and fury!" he roared. "What kind of bread is this? Look here, I've lost two teeth trying to eat it!"

"I am terribly sorry," whimpered Oonagh. "I didn't realize my bread would be too hard for your tender mouth. You see, this is the only kind my Fin and his baby will eat, but then, they have very strong jaws."

"Well, if Fin can eat this, I can, too," bellowed Cucullin. "Give me another loaf."

Oonagh did as requested, but after one chomp, Cucullin shouted, "Thunder and giblets, I've lost two more teeth! Does Fin actually eat this bread?"

"Why, yes, and so does our baby. You must stop shouting, though, or you will wake the little one."

When Fin heard Oonagh's words, he cried out in a high squeaky voice, "Mama, I'm hungry. I want bread. Give me bread."

Carefully Oonagh selected the one loaf without a griddle and gave it to Fin. Cucullin tiptoed over to watch. He was amazed at the size of the child and was shocked when the baby gobbled the bread without a roar or a shout.

"My," thought Cucullin, "Fin must be a monster of a giant to have such a son."

Just then, Fin looked up at Cucullin and said, "Are you strong—as strong as I am? I can make water come out of a stone. Can you?"

"Junior, stop bragging. Certainly you can get water from a stone. Your papa could, too, at your age. Surely Mr. Cucullin dribbled water from rocks when he was a tot," said Oonagh. Then she handed Cucullin a white stone and asked, "Could you amuse the child by squeezing out a few drops?"

"Hmmmmm," muttered Cucullin, "water from a stone. Well, if Fin's baby can do it, I can, too."

With these words, Cucullin took the stone and squeezed. But no liquid emerged.

"You call yourself strong?" laughed Fin. "Look, this is how you do it." Fin sat up and grabbed the stone but then quickly, slyly, switched it for a handful of curds. He squeezed the curds and drip, drip, drip—out came clear water.

Now Cucullin was scared—really scared. He did not want to fight a giant who picks up his house when the wind blows, digs through solid rock without breaking a sweat, eats bread hard enough to break teeth, and has a baby who eats the same bread and can squeeze water from a rock. No indeed, Cucullin did not care to battle Fin. The only thing he wanted was to get as far away from Knockmany Mountain as he possibly could.

And so he said, "Mrs. M'Coul, thank you for your hospitality, but I must be leaving now."

"I understand," said Oonagh.

"There's just one thing before I go," said the giant. "Would you let me look into your baby's mouth? I'd like to see the teeth that can eat your bread."

"Of course," said Oonagh.

And so Fin opened his mouth and in went Cucullin's magic finger. When that happened, Fin knew just what to do. He closed his mouth and chewed. Fin gnawed so hard, he tore the finger off Cucullin's hand.

"The baby's stolen my strength!" Cucullin wailed. Then, squealing in rage and fear, he ran away from the house and down the mountain. You can be sure he never bothered Fin M'Coul or clever Oonagh again. And so Fin, freed from fearing Cucullin, and Oonagh, with her house turned exactly as she had always wanted it and with an ever-so-convenient new well, lived happily ever after.

Notes

About the Story

In this tale, a mighty giant is defeated by a physically weaker but much smarter opponent. Typical fairy-tale plot, right? Only in this case, the weaker opponent is not a courageous girl or valiant lad. Here, the weaker character is a giantess. The humor of this situation will not be lost on young listeners.

Before starting the story, you might want to describe curds and whey to your students.

To retell the legend, I used two sources:

Phelps, Ethel Johnston. *Tatterhood and Other Tales*. New York: Feminist Press, 1978, pages 87–92.

Yeats, W. B. *Fairy and Folk Tales of Ireland*. New York: Collier Books, Macmillan Publishing Co., 1973 (from a text first published in 1888 and 1892), pages 241–251.

Your Thoughts

The Shrinking Giant

The students' goal in this activity is to find Cucullin's final height after he has undergone 15 days of shrinking. They can keep track of the shrinking by recording the giant's daily decrease in height. In the process, they will get lots of practice in converting measurements of inches and feet.

Over the 15-day period, the giant goes from a height of 12 ft to a height of 9 in. Some children may calculate the shrinking with paper and pencil. Others may start with a 12-ft-long piece of string and then measure and cut off the required number of inches for each day's shrinking. Done carefully, the length of the resulting piece of string should be equivalent to the giant's final height.

You may want to begin this activity by stretching a 12-ft-long piece of string across the floor so children can visualize the giant's size. It will be fun for children to compare that great height with their 9-in.-high drawings.

When Cucullin ran away from Knockmany Mountain, he was not happy. No, not at all. But he was even more miserable the next day. Why? He noticed that his clothes were all too big.

"Oh, no," he thought, "I'm shrinking."

It was true. Instead of his normal 12-foot height, Cucullin was now a mere 11 feet, 10 inches. He had lost 2 inches of height in one day! The next day he lost 3 more inches! And the day after that he was 4 inches shorter! For 15 days he got smaller and smaller—always shrinking one more inch each day than the last.

Then, on the 15th day, the shrinking stopped.

Tell how tall Cucullin was at the end of each shrinking day. How tall was he when he stopped shrinking?

Draw a picture of the short Cucullin. Make sure the drawing shows him at his exact height after 15 days of shrinking.

The Shrinking Giant

The day after Cucullin ran away from Knockmany Mountain, he made a horrible discovery.

He was shrinking!

Scared as could be, he measured himself. Normally he was 12 feet tall, but today he was 2 inches shorter—a mere 11 feet, 10 inches.

The next day, he lost 3 inches *more*. The day after that, he lost 4 inches.

Every day he lost one more inch than he had the day before.

This lasted for 15 days.

Tell how tall Cucullin was at the end of each day.

How tall was he when the shrinking stopped?

Draw a picture of the new Cucullin.

Make sure it shows him at his exact height.

Six Colors

There are 10 ways to arrange six colors into three sets of three colors each. If children create models of the threads using strips of colored paper, they can manipulate the strips and record results.

Here are the 10 combinations using initials: RBY/GOP, RBG/YOP, RBO/GYP, RBP/GOY, RYG/BOP, RYP/GBO, RYP/GOB, RGO/BYP, RGP/BYO, ROP/GYB.

Two weeks after Cucullin ran away from Knockmany mountain, Oonagh had another problem to solve. Fin's birthday was just three days away and she did not know what to give him.

Oonagh went to her weaving shed. There she selected six strands of yarn in six different colors: red, blue, yellow, green, orange, and purple. She decided that two braids, each with three strands of yarn, would provide enough bright thoughts for her to solve this problem.

Using all of her six strands of different colors Oonagh could have created a red-blue-yellow braid and a green-orange-purple one. That's two braids, each having exactly three colors. Another arrangement she could have created is red-blue-green and yellow-orange-purple.

There are eight more ways she could have arranged six colors into two braids. Your job is to name them all.

Which combination do you think is the brightest?

What gift should Oonagh give Fin?

Six Colors

Oonagh needed to give Fin a birthday present, but she did not know what to give her dear husband. To help her have bright thoughts Oonagh went to her weaving shed and took six strands of yarn in six colors: red, blue, yellow, green, orange, and purple.

Using these six colors she made two braids. Each braid had three strands of yarn.

Oonagh could have created a red-blue-yellow braid, and a green-orange-purple one.

Or she could have created a red-blue-green braid and a yellow-orange-purple one.

There are eight more ways she could have arranged six colors into two braids. Your job is to name them all.

Which combination do you think is the brightest?

What gift should Oonagh give Fin?

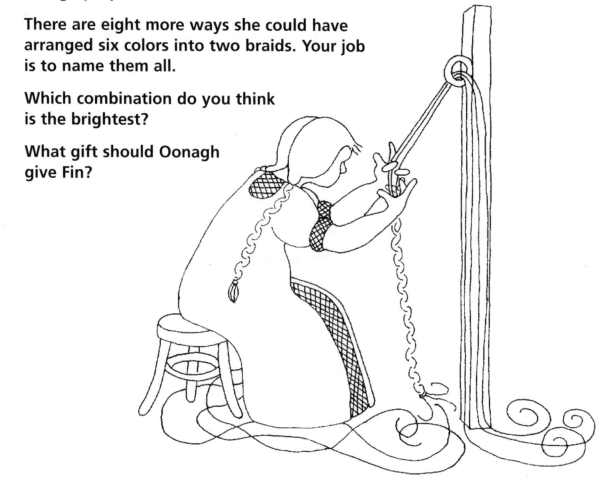

Oonagh's Necklace

Most students will use trial-and-error to discover the correct number of copper, brass, silver, and gold links in Oonagh's necklace. Counting materials such as counting cubes can aid students in their work. After discovering that there are 9 copper, 18 brass, 27 silver, and 36 gold links, children will be ready to design their own necklaces. You might encourage them to draw patterned designs for their creations.

Fin knows how lucky he is to have a wife like Oonagh. To show his admiration and gratitude, Fin decides to make her a beautiful necklace.

He has a collection of copper, brass, silver, and gold links all ready to form into a chain.

Find out how many links of each kind Fin has for Oonagh's necklace.

Here are your clues.

1. Fin has 90 links in all.

2. He has twice as many brass links as copper ones.

3. He has three times as many silver links as copper ones.

4. He has four times as many gold links as copper ones.

Using Fin's links, how would you design Oonagh's necklace?

Oonagh's Necklace

Fin decides to make a thank-you necklace for Oonagh.

He has a collection of copper, brass, silver, and gold links ready to form into a chain.

Figure out how many links of each kind Fin has.

Here are your clues.

1. Fin has 90 links in all.

2. He has twice as many brass links as copper ones.

3. He has three times as many silver links as copper ones.

4. He has four times as many gold links as copper ones.

Using Fin's links, how would you design Oonagh's necklace?

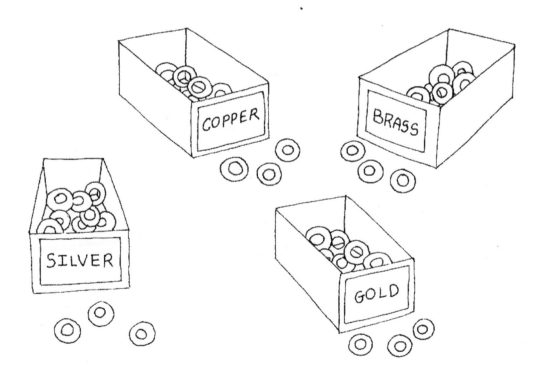

Steps Saved

To figure out the number of meters Oonagh saves in a day, a week, a month, and a year, students can multiply, perhaps with the help of a calculator. Less formal methods—adding, counting, manipulating base-10 materials, or drawing—can also lead to accurate solutions.

You may choose to make this problem easier by reducing the length of the path down the mountain or by having students merely find the number of meters saved in one day.

Oonagh saves 8 km (8,000 m) a day; 56 km (56,000 m) a week; 248 km (248,000 m) in May; 240 kilometers (240,000 m) in June; and 2,920 km (2,920,000 m) a year.

Oonagh loves her new water well. Now she can fill her buckets without climbing up and down Knockmany Mountain. That's a giant relief for her. After all, the path from her house down to the base of the mountain is 800 meters long and Oonagh has to fetch water five times a day.

Can you figure out how many meters of walking Oonagh saves each day thanks to Cucullin's digging powers?

How many meters of walking does Oonagh save each week? How many during the month of May? How about the month of June? How many meters does she save each year?

Steps Saved

Oonagh loves her new well. Now she does not have to climb up and down Knockmany Mountain five times a day to fetch water.

The path from her house down to the base of the mountain is 800 meters long.

How many meters of walking does Oonagh save each day?

How many meters does she save each week?

How many during the month of May?

How about the month of June?

How many meters does she save each year?

> No more down and up down and up. Isn't it wonderful!

A Tale from Russia

Once upon a time a widowed merchant and his daughter, Lubachka, lived in a lovely cottage near the woods. Father and daughter were happy together, at least they were until the merchant decided to remarry.

For a short time—not a long time—the merchant's new wife was kind both to her husband and her stepdaughter. But then everything changed. You see, the merchant often traveled from town to town buying and selling goods. The very first time he left his new wife alone with his child, the woman began treating the girl in a cruel, demanding way. She forced Lubachka to mop floors, scrub walls, scour pots and pans, cook meals, mend clothes, make beds, collect firewood, and care for the pigs, chickens, ducks, and cows.

Lubachka worked from dawn to dark. She did not stop for breakfast or lunch. At dinner-time, her stepmother tossed Lubachka a few crusts of stale bread and sent her to the woodshed to eat.

Lubachka's father did not know how his daughter suffered during his many absences. You see, whenever he returned from his travels, Lubachka's stepmother treated the girl tenderly, oh, so tenderly. But the moment he went away again, everything changed.

One evening, as Lubachka crouched in the woodshed nibbling crusts, she heard a scratch, scratch, scratching sound. What was making that noise? A little gray mouse hunting for food. Feeling sorry for the hungry animal, Lubachka offered him her handful of bread

and he gobbled it up.

"Thank you, thank you. You are so kind," the mouse whispered as he scampered away.

The next night, the same gray mouse was waiting for Lubachka by the woodshed door. And although she only had a single crust of bread, Lubachka split this meager meal with her new friend. Night after night Lubachka shared her food with the mouse. One evening, the little creature gave his benefactor a warning.

"Your stepmother has a sister," he began. "This sister is the bony-legged one, witch-of-the-woods, Baba Yaga by name. Someday your stepmother will ask you to visit her sister. When she does, come quickly to me. For Baba Yaga will surely eat you with her iron teeth unless you follow my instructions."

A short time—not a long time—after the mouse's warning, Lubachka's stepmother announced, "Today you will go to my sister's home deep in the woods. When you arrive, ask for a needle and thread and then bring them back to me."

Lubachka did not argue with her step-mother. But, before leaving, she wrapped a few crusts of bread in an old towel and went to the shed to visit the mouse.

After eating, the mouse said, "Now I will tell you all you need to know. Carry this towel with you. It will be very useful. Also, as you follow the trail to Baba Yaga's house, pick up and save every odd thing you discover. If you do this and keep your heart kind, you will defeat the iron-toothed witch."

Lubachka thanked the mouse and started walking through the woods. She had not gone far when she saw a bright red handkerchief dangling from a thorn bush. Lubachka grabbed it and tucked it into the waistband of her apron. Next, she found a loaf of bread and some slices of roast ham resting on a rock. The food smelled extraordinarily tasty and Lubachka was very hungry but, remembering the mouse's words, she carefully wrapped the bread and ham in her towel. Near the rock, she found a wooden hair comb and a can full of oil. She dropped these into the pocket of her jacket.

After a short time—not a long time—Lubachka reached Baba Yaga's home. She walked up to the gate and pushed it open. The hinges were extremely rusty and very squeaky. Indeed, they were in such dreadful condition that kindhearted Lubachka felt sorry for the gate. Without hesitating, she poured oil from her oil can all over the hinges. The gate seemed to sigh in relief.

A moment later, Lubachka saw a housemaid weeping next to a tall oak tree.

"Why are you crying?" asked Lubachka.

"Do you see that giant mortar and pestle in the middle of the yard?" asked the maid. "Baba Yaga, the bony-legged witch, sits in that mortar and flies from here to there steering with the pestle. This morning she left her flying contraptions outdoors. She ordered me to carry them inside, but I was not strong enough to lift either mortar or pestle and so she beat me with a broom."

"You poor girl," said Lubachka, "here, take this red handkerchief and wipe your eyes."

Leaving the housemaid, Lubachka walked up to Baba Yaga's door and knocked.

"Who is at my door?" growled the bony-legged one through her shiny iron teeth.

Lubachka answered, "Good day, Auntie. I've come from your sister, my stepmother. She wants to borrow a needle and thread."

"My dear, my beloved," said Baba Yaga using a sugary voice, "come into my home. Now my darling, my little one, sit here and weave on my loom until I find what you need."

After Lubachka started weaving, Baba Yaga walked out of the house and into the yard. She went to her maid and snapped, "Fix a bath. Make it hot. Then scrub, scrub, scrub my little niece. I want her very clean for I plan to eat her for dinner."

Through the open window, Lubachka heard Baba Yaga's words. And so, when the maid came through the house carrying a bucket of water, Lubachka begged, "Please be slow in filling the tub. Use a sieve, not a bucket. Be slow in heating the water. Wet the wood so that it will not burn."

The housemaid said nothing. She was too afraid of Baba Yaga to talk, but she did smile.

Lubachka turned back to the loom. A short time—not a long time—later, Baba Yaga called out, "Are you weaving, my love?"

"Yes," answered Lubachka, "I am weaving, Auntie."

As Lubachka worked, she noticed a cat—oh, such a skinny little cat—staring at her. Lubachka felt profound pity for this tiny creature. So she reached into her towel and pulled out the slices of ham. "Here, Catakins," she called, "eat these."

The cat devoured the meat and then whispered, "Lubachka, Lubachka, you must get away from this house. If you stay, Baba Yaga will cook you and eat you for her dinner."

"How can I escaped a witch? She can fly. I cannot," said the girl.

"Run. Run fast through the woods. True, when Baba Yaga discovers you are gone, she will chase you in her flying mortar. Wait until she gets so close that you can see her red eyes. Then toss that towel behind you. The towel will turn into a wide river. The mortar cannot fly over water. Of course, Baba Yaga will find a way to overcome this obstacle, and she will be after you again. Wait, wait until you feel her breathe on your heels. Then toss your comb on the ground behind you. The comb will turn into a dense forest. Baba Yaga will not be able to pass through such a mass of trees."

"Thank you, dear Catakins, for your help," said Lubachka.

Immediately, Lubachka left the house and started running. Halfway across the yard a huge dog growled menacingly at her. Instantly, Lubachka threw him her loaf of bread, and the big animal started eating.

So Lubachka, still clutching the towel and the comb, ran through the gate and took off into the woods.

After a short time—not a long time—Baba Yaga called from the back of her house, "Lubachka, my beloved, are you still weaving at my loom?"

"Yes, Auntie," the skinny cat hissed, trying to sound as girl-like as he could.

"That is not Lubachka's voice! Who is at my loom?" screeched Baba Yaga.

The witch rushed into the room. What did she see? Her cat rolling about, mixing up all the threads.

"Lubachka is gone," announced the cat.

"Cat, oh Cat," shouted Baba Yaga. "Why didn't you stop her?"

"Baba Yaga, you never feed me—not so much as a bit of cheese. While Lubachka, the kindhearted, gave me savory slices of ham!" said the cat.

Then Baba Yaga called her maid. "Why didn't *you* stop Lubachka?" the witch screamed.

"Baba Yaga, you beat me and make me cry, while Lubachka, the sweet, gave me a red handkerchief to wipe my tears," answered the maid.

Baba Yaga ran into the yard and yelled, "Dog, why didn't *you* stop Lubachka? You could have bitten her."

"From you, Baba Yaga, I never get a crust of bread. Lubachka, the good, gave me a whole loaf!" replied the dog.

"Gate, why didn't *you* creak and squeak? That would have warned me of Lubachka's escape," yelled Baba Yaga.

"My hinges ached and but you did not care. Lubachka, the generous, gave me healing oil!" said the gate.

"You have all betrayed me," raged Baba Yaga. "But I will catch Lubachka. I will catch her and bring her back for my dinner."

Then Baba Yaga grabbed her pestle, jumped into her mortar, and began flying.

In a short time—not a long time—Lubachka, running between the trees, heard Baba Yaga's cruel laugh. Swirling around, Lubachka saw Baba Yaga's blood-red eyes. Quickly, quickly, Lubachka threw her towel onto the ground. Suddenly, a wide river streaked across the woods. Baba Yaga's mortar came to a halt. It could not fly over the water.

Baba Yaga was angry. Oh yes, she was. But she would not let this river keep her from her dinner. She flew back to her barn, got her team of oxen, led them to the river, and set them to drinking. Soon, the oxen had swallowed every drop of water, making the river disappear. Then Baba Yaga climbed into her mortar and started flying again.

Lubachka heard the witch, saw the witch, felt the witch's breath on her heels. Quickly, quickly, Lubachka threw down her comb, and up rose a dense, dark forest. Baba Yaga was trapped behind a web of trees.

What did Lubachka do then? She ran all the way home. There she saw her father. Immediately, she told him everything that had happened. Learning of his daughter's suffering, the merchant roared in anger. Lubachka's stepmother heard the roar and, fearing her husband's fury, ran off into the woods. Where did she go? No one knows. But she never returned. As for Lubachka and her father, for a *long* time—not a *short* time— they lived happily ever after.

Notes

About the Story

Before reading the story, you might want to explain that a mortar is a bowl- or cup-shaped receptacle in which grain, seeds, and other materials are ground into powder. A pestle is a club-shaped tool that is used to pound the grain in the mortar.

Baba Yaga is the evil protagonist of many Russian folk tales. With a little hunting you can find more stories about "the iron-toothed, bony-legged one" to share with your class.

For my retelling, I relied on three versions of the story.

Afanas'ev, Aleksandr; trans. Guterman, Norbert. *Russian Fairy Tales*. New York: Pantheon Books, 1945, 363-365.

Ransome, Arthur. *Old Peter's Russian Tales*. New York: Dover Publications, 1969, 90-105.

Wyndham, Lee. *Tales the People Tell in Russia*. New York: Julian Messner, 76-87.

Your Thoughts

Baba Yaga's Spell Book

To solve the spell book's problem, children must discover ten out of the twelve factors of 96. A calculator will prove helpful. Youngsters who understand factors will realize that each time they find one number that divides 96 evenly, they find a second factor as well—namely the whole-number answer to the calculated division problem. You can, of course, change the difficulty of the problem. Make it harder by demanding that children find all twelve factors. Make it easier by requesting fewer factors. The factors of 96 are: 1, 2, 3, 4, 6, 8, 12, 16, 24, 32, 48, 96.

When Baba Yaga returned home, she discovered that her housemaid, cat, and dog had run away. Angry—this made Baba Yaga very angry. She wanted to bring the trio back, and she wanted to bring Lubachka back, too. A proper spell—that's what she needed. A spell to shrink all the trees Lubachka's comb had created until they became itty bitty leaves of grass.

Baba Yaga fetched her favorite spell book. A hundred years ago, she had stolen this book from a wizard. While he owned the book, the wizard did something very clever to protect his spells. He put a magical lock on the book. From that time on, whenever anyone lifts the spell book, a mathematical question appears on the cover. The book will not open unless this question is answered correctly. The book never asks the same question twice. And if the question goes unanswered or is answered incorrectly, the book will not not produce another question for one week.

The book's math problems never troubled the wizard. He was a great mathematician. Baba Yaga, however, rarely knows the right answer. That does not stop her from trying, though.

After fetching the book, Baba Yaga looked on the cover. There she read, "There are just twelve numbers that divide 96 evenly. To open me, you must name at least ten of these numbers."

What numbers did Baba Yaga have to name? Would you be able to answer the question and open the book?

Baba Yaga
Mouse House

While designing their mouse houses, children get some architectural experience with area and perimeter. If you make a transparency of the grid for an overhead projector, you can build a sample house with your class. Children can brainstorm ideas for the five rooms. Suggest various ways to shape the rooms. Make sure, though, to reinforce the importance of keeping individual square inches intact by drawing only along the grid lines and not through them. Unless students do this, they cannot accurately calculate areas and perimeters. The total area of each mouse house will, of course, be the same for all students: 225 square inches. The total perimeters will vary.

Lubachka was very grateful to her friend, the mouse. After all, without his advice, she might have been Baba Yaga's evening stew. So, as a gift, Lubachka planned to make a lovely house for the mouse. She decided the building would take up a 15-by-15-inch section of the woodshed.

Lubachka wants to divide the house into five separate rooms. It is hard, though, to select the right rooms. The mouse may like a living room, a cheese storage room, a guest room, or a playroom. He may enjoy a bedroom, a bathroom, a kitchen, or a singing room. Lubachka cannot make up her mind.

If it were your job, how would you design the mouse house? Use the Mouse House grid to show your ideas. First, select five rooms. Then draw them on the grid and label them. You can lay out the rooms in any shape you like—L-shape, T-shape, square, or rectangular—but you must draw the walls along the grid lines.

When you finish your plan, figure out how many square inches are in each room. Then add all the square inches. How many do you get? Compare this number with your classmates' numbers. Do they get the same total area as you do?

Now figure out the perimeter of each room. Add all the perimeters. Compare this number with your classmates' numbers. Do they get the same total perimeter as you did?

Mouse House

Here is a floor plan for the mouse house.

Draw five rooms and label them.
Find the area and the perimeter
of each room. Add all the areas.
Then add all the perimeters.
Compare your results with your
classmates' results.

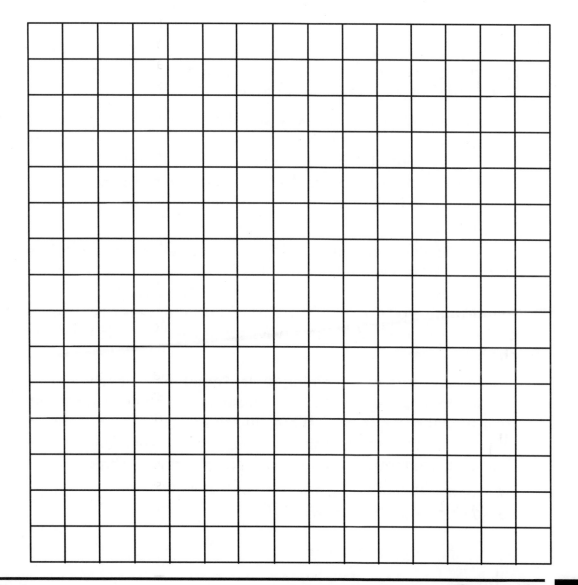

Baba Yaga

The Dinner Table

This is a permutation problem. To discover all the seating arrangements, children can draw pictures, use trial and error, or create an organized list. Encourage children to find as many of the 22 missing seating arrangements as possible, but don't worry if their lists fall short.

The place arrangements (using initials to represent the Housemaid, Cat, Dog, and Mouse) are: HCDM, HCMD, HDCM, HDMC, HMCD, HMDC, MCDH, MCHD, MDCH, MDHC, MHCD, MHDC, CMHD, CMDH, CHMD, CHDM, CDMH, CDHM, DHMC, DHCM, DCMH, DCHM, DMCH, DMHC. If children have trouble approaching this activity, show them how to keep track of the arrangements in a chart like this one.

Seating Arrangements	Chairs			
	1	2	3	4
1.	H	C	D	M
2.	H	C	M	D
3.				

"We must run away from here," said the cat as soon as Baba Yaga left to chase Lubachka. The maid and dog agreed. The gate wanted to join them, but, of course, he could not.

The escaping trio ran this way and that. They got lost a few times, but eventually they left the trees behind. Soon they found themselves standing in front of a lovely cottage. To their surprise, there was Lubachka at the front door. The kindhearted girl was delighted to see her friends. So delighted, she invited them to live with her and her father.

Every evening, Lubachka sets six places at her dinner table. There are places for Lubachka and her father, and for the housemaid, the cat, and the dog. There is also a place for the little mouse who had done so much to help Lubachka. Lubachka's father always sits at the head of the table while Lubachka always sits at the foot. The four others do not have permanent places.

Here's a picture of the dining table. Sometimes Lubachka's friends position themselves like this: the housemaid in chair one, the cat in chair two, the dog in chair three, and the mouse in chair four. At other meals, they sit like this: the housemaid in chair one, the cat in chair two, the mouse in chair three, and the dog in chair four.

There are 22 more ways for the cat, the dog, the mouse, and the housemaid to sit. How many of them can you list?

The Dinner Table

Together the housemaid, the dog, and the cat ran away from Baba Yaga's. They went to live with Lubachka and her father.

Every night Lubachka sets six places at the dinner table. Her father sits at the head of the table. Lubachka sits at the foot. The housemaid, the cat, the dog, and the little mouse who lived in Lubachka's woodshed sit in the four other places. But these four keep changing seats.

There are twenty-four ways that Lubachka's four friends can arrange themselves at the table. How many of the twenty-four can you list?

Spell Away

As children figure out how many centiliters of each ingredient to include in their spell repellents, they will do a considerable amount of addition and subtraction. There are, of course, a vast number of workable combinations. Children will get some sense of this as they compare their lists. This activity makes a particularly good homework assignment. When children are finished, they can write additional magic formulas. How about a strength-building formula or a flying formula? Some students may want to publish a book filled with their concoctions.

After she arrived at her home, Lubachka told her friend the mouse how she had escaped from the witch. The mouse was impressed by Lubachka's courage, but he knew that Baba Yaga would not give up. The iron-toothed one would surely cast some kind of spell that would get her through the woods. Indeed, she would cast spell after spell until she got Lubachka back.

How did the mouse know so much? Before moving into Lubachka's woodshed, he had occupied a hole in Baba Yaga's living room wall. There he learned the ways of the witch. Better yet, he learned to make a special spell repellent—a liquid formula that protects the wearer from witches' spells.

The mouse explained how to make the repellent. "Lubachka," he said, "you must brew exactly one liter of this special formula. That is the same as 100 centiliters. All the ingredients must be in liquid form. There are many liquids you can use, but you must use six or more different ones." Then the mouse gave Lubachka a list of possible ingredients.

Using the mouse's instructions and list of ingredients, you can write your own formula for a spell repellent. Pick the liquids you think would be most effective against a witch's spell. You may add to the list if you like. Decide how many centiliters of each liquid to use. Why did you select each of your ingredients? Why do you think your mixture would make a good spell repellent?

Baba Yaga

Spell Away

The mouse told Lubachka how to make a spell repellent—a liquid that can protect her from Baba Yaga.

Using the mouse's instructions and list of ingredients, you can write your own formula for a spell repellent. Pick six or more ingredients—ones you think will be most effective against a witch's spell. You may add to the list if you like. Decide how many centiliters of each liquid to use. You must end up with exactly 100 centiliters of repellent, no more, no less.

Ingredients List

poison ivy sap	dog drool
bat's blood	perfume
glue	honey
spit	bug juice
lemon juice	liver soup
green slime	baby tears
gasoline	hot chocolate
rusty water	vinegar
muddy water	rotten milk
dishwashing liquid	shampoo

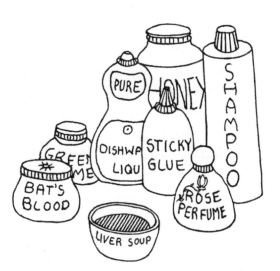

Why did you select each of your ingredients? Why do you think your mixture would make a good spell repellent?

The Ebony Horse

A Tale from 1001 Arabian Nights

Once upon a time the Sultan of Persia and his son, Prince Kamar al-Akmár, were enjoying themselves at the annual winter celebration when they were approached by a strange little man leading a most unusual horse. This animal was made entirely of wood—dark ebony wood.

"Sire," said the man, "allow me to entertain you with my horse—my flying horse. It is true, this remarkable creature can soar through the skies."

"Impossible!" shouted the sultan, and his courtiers nodded their agreement.

But the sultan was wrong—it was possible. As everyone watched, the little man mounted the horse and began gliding through the air.

The sultan was amazed, astonished, delighted. "I must own this phenomenal creation. What can I give you in exchange for it?" he asked.

"Your daughter," said the little man, "I want to marry your daughter."

"Clearly, you are a clever man, a great magician," said the sultan, "but you cannot marry my daughter. Only one of royal blood can spend his life with her. I will, however, select an appropriate payment for you. First, though, I want my son to learn how to operate this extraordinary animal."

So Prince Kamar al-Akmár mounted the horse. The magician touched a small peg on the horse's neck and said, "Pull on this to make the horse fly."

Without waiting for further instructions, the prince tugged on the peg. Instantly, the beast rose into the air. Moments later, horse and rider disappeared from view. The sultan smiled as he watched his son sail away.

But then the magician, shaking and quivering, cried out, "Oh, great and powerful one, the prince does not know how to land the horse. He flew off before I could explain."

Imagine the sultan's horror. His son—his beloved son—was lost in the skies on a magic horse without enough knowledge to descend to the earth! The magician must be held accountable for this catastrophe. "Guards," shouted the sultan, "take this horrible little man to prison and keep him there until my son returns."

The prince, meanwhile, riding on air, was completely happy. What a delight to float over villages, towns—even kingdoms. After several hours, though, he grew tired and hungry. He wanted to land. The prince tried this and that until he discovered a second peg. He pulled on it and the horse coasted downward.

Where did the prince land? In a forest? In a meadow? In the middle of the ocean? No. The horse came to rest on a terrace near the top of a magnificent palace. Looking through the open terrace doors, he saw a splendid bed chamber. Stepping inside, he found himself staring at a beautiful princess.

The princess was shocked to see a strange man in her room. She was about to summon her servants when the prince begged her to listen to his story. She agreed and was spell-bound by his amazing tale. Then she did summon her servants and ordered them to prepare an enormous supper for her surprise visitor.

The next morning, the princess invited the prince to remain in the palace as an honored guest, and he was happy to oblige.

Day after day, the prince enjoyed exquisite food and charming entertainments in the company of the princess and her court. Soon the prince and the princess fell in love and decided to marry. They agreed to fly the ebony horse to Persia and inform the sultan of their plans.

When they reached Persia, the prince flew the princess to a summerhouse near the main palace. He left her there with the horse while he went to see his father.

The sultan was overjoyed to be reunited with his son. Immediately he ordered a grand feast to celebrate the prince's homecoming and to welcome his future daughter-in-law. Then, true to his word, he ordered the magician's release from jail.

Naturally, the magician was pleased to leave prison. But did he forgive the sultan for locking him away? Did he forgive the prince for flying off so impetuously? No, he did not. Indeed, he was furious with both the sultan and Prince Kamar al-Akmár.

As he departed his jail cell, the magician overheard several guards talking. They were discussing the prince, his return to Persia, the ebony horse, and the princess. Using this information, the magician devised a plan—an evil plan—a plan of revenge.

As soon as he was free, the magician went directly to the summerhouse and presented himself to the princess. "My lady," he said, "Prince Kamar al-Akmár requests your presence at the palace. I am to escort you. We shall fly there on the ebony horse."

Did the princess believe the magician? Unfortunately, she did. She eagerly joined him on the horse and they took off. Almost immediately, though, the princess noticed that they were heading away from the palace and not toward it. She asked the magician for an explanation.

The magician laughed and said, "Princess, you are at my mercy. The sultan would not let me marry his daughter. Fine, now I have you instead, and I shall take you so far from Persia that the prince will never find you!"

After several hours, magician and princess were flying over the Kingdom of Cashmere. The magician, being tired, set the horse down in a lush meadow. Within moments, he was snoring loudly. Miserable and angry, the princess could not sleep. She was wide awake, therefore, when a troop of horsemen galloped into sight. She called out to their leader.

"Who are you?" asked the leader, "What brings you to my meadow?"

At that moment, the magician woke up, and before the princess could reply, the magician began talking. "My wife and I are travelers in your land. Exhausted from our journey, we fell asleep in your field."

"He lies," interrupted the princess. "I am this man's prisoner, not his wife. I beg you, kind sir, help me escape."

"Dear lady," proclaimed the leader, "it happens that I am king of this land. As such, I am happy to offer you my complete protection."

The Ebony Horse

Then the King of Cashmere escorted the princess to his palace, while his guards dragged the angry magician to jail. What happened to the horse? The king's soldiers moved it to the palace courtyard, for, although the king did not know that the horse could fly, he did value its beauty.

As the princess entered the palace, she felt sure that soon she would be reunited with Prince Kamar al-Akmár. But it was not to be. You see, the king was so impressed by princess's bravery, intelligence, and beauty that he fell in love with her and declared that he wanted to marry her. Of course, the princess refused the king's proposal, but that did not discourage him. He was certain that after a majestic wedding the princess would come to love him as he loved her. With this in mind, he sent her off with his servants. Then he began planning a spectacular ceremony.

The princess knew that crying, begging, and whining would not help her get away. So she thought up a clever plan instead.

The next morning, when the king's servants entered the princess' bed chamber, they were greeted by an odd sight. There was the princess flapping her arms and making cooing sounds. Why she was behaving exactly like a bird! At lunchtime the princess growled like a lion and insisted on eating raw meat off the floor. That evening she shouted nonsensically throughout a royal concert.

Every day, all day, the princess either behaved like an animal or talked in a thoroughly bizarre manner. Clearly she was mad, crazy, out of her mind. The king summoned the palace physician. Screaming wildly, the princess smashed the physician's tray of powders and potions. The king sent for doctors from neighboring towns and kingdoms. But, when they arrived, the princess yowled and kicked so ferociously that they all ran away. The truth was that only the sight of Prince Kamar al-Akmár could cure the princess.

Where was the Persian prince all this time? He was searching and searching for his missing sweetheart. You see, when he discovered that the princess, the horse, and the magician had all disappeared, he guessed what had happened. That very afternoon the prince set out to rescue his true love. Before leaving the palace, however, he traded his royal garments for beggar's rags. Disguised as a beggar, he could ask questions, overhear conversations, and collect information without arousing suspicion.

The prince wandered from land to land, traveling on and on until he reached the Kingdom of Cashmere. There he heard stories of a life-size horse made of ebony, a princess who had lost her wits, and a king who would do anything to cure her.

Cheered by this news, the prince went to the palace and pleaded for an audience with the king. The guards laughed at this poor fellow dressed in rags.

"Please," insisted the prince, "go to your king. Tell him that I come from a distant country and have the power to cure many unusual maladies."

Hearing these words, the guards stopped laughing and led Prince Kamar al-Akmár directly to the king. The king escorted him to the princess. They found her in an elegant sitting room bellowing like a cow. The moment she saw Prince Kamar al-Akmár, though, the princess ceased her outrageous behavior. She stood quietly and smiled.

Smiling back at her, the prince said, "Your highness, I am a healer from a foreign land. I hope you will let me help you."

"I will and gladly," replied the princess.

The King of Cashmere was amazed. These were the first reasonable words the princess had uttered in weeks. For the next few minutes the princess spoke quite sensibly to her make-believe doctor. The king was delighted. It appeared that his princess was cured.

"Your majesty," said Prince Kamar al-Akmár, "this is only a temporary remedy. To permanently free the princess of her madness, I must know what enchanted her. Did she arrive here with any unusual objects?"

"Yes," said the king, "she arrived with an ebony horse. It is in my courtyard."

"Excellent. In that case, I promise a complete recovery," said the prince. "But first, I must spend time alone with the princess and this horse."

The king led the prince and the princess to the courtyard. He left them there and returned to his throne room. Moments later, he looked out the window. What did he see? The prince and princess flying through the skies on the ebony horse. The king called out to the princess, but she could not hear his pleas.

In a few hours, the prince and princess reached Persia. They were married that very evening. It was the most spectacular wedding in the history of that very great land. The celebration lasted two weeks. From that time on, the prince and the princess lived happily ever after.

Notes

About the Story

The stories in *1001 Arabian Nights* come from many different lands, such as Persia, India, and Egypt. Some stories probably originated in China and Japan.

In the early 1400s, Egyptian scholars recorded the stories. Copies of these made their way around the Arab world. In 1704, Antoine Galland, a French scholar, discovered the stories in Syria. He began translating them into French but never completed the job. Even so, his translations introduced the stories to Europeans.

More than 150 years later, Sir Richard Francis Burton discovered the tales and translated them into English. In an effort to explain Muslim culture and customs to Western readers he added copious footnotes. The Heritage Press edition of his work runs more than 3,870 pages.

For this retelling, I relied on four versions of the story.

Burton, Richard, F. *The Book of The Thousand Nights and a Night*. New York: Heritage Press, 1934, Vol. 3 and 4, 1567–1598.

French, Marion N. *Myths and Legends of the Ages*. New York: Hart Publishing Company, 1951, 245–254.

Lang, Andrew, ed. *The Arabian Nights Entertainments: Aladdin, Sindbad and 24 Other Favorite Stories*. New York: Dover, 1969, 358–389.

The Arabian Nights. New York: Grosset & Dunlap, 1946, 62–78.

Your Thoughts

Everyone Wants to Fly

To figure out when the Prince gets his second turn to ride, students can plot the daily riding schedule rider by rider. The Prince's first ride is from 11:00 to 11:45. After the horse rests for 10 minutes, the next person rides from 11:55 to 12:40 and the horse rests again. By repeating this process twice more to reach 2:30, students discover that four people can ride each day. Now they must account for 14 riders—4 on Monday, 4 on Tuesday, 4 more Wednesday—that's 12. On Thursday, after the remaining two riders get their first turns, the prince will get his second turn. The time? 12:50.

Shortly after their wedding, Prince Kamar al-Akmár taught the princess how to operate the ebony horse. He also taught his father, his mother, his five brothers, and his five sisters. After that, sad to say, the family was always fighting. Why? Because everyone wanted to ride at the same time.

Prince Kamar al-Akmár hated to hear his family argue, and so he devised a plan to please everyone. Here is his plan: They will all take turns. Each rider will get the horse for exactly 45 minutes. After each ride, the the horse will get a 10-minute rest. Then the next person will get a turn.

The royal family will ride every day from 11:00 AM until 2:30 PM. If Prince Kamar al-Akmár takes his first turn on Monday at 11:00 AM, when will he get his second turn?

Do you think the prince's plan is fair? Can you think of another plan?

The Ebony Horse

Getting Out of Jail

Students will be challenged by this two-step problem. First they must find out how many days the magician spent in jail. To do this, they must divide 270 (total meals) by 3 (meals a day) to get an answer of 90 days. They could certainly obtain the same results by using repeated subtraction, repeated addition, trial-and-error, or by modeling the problem with counting materials.

Next they have to determine how much money the magician owes for food. Again, they can multiply—90 (days) times $5.00 (cost per day) for a result of $450. But other methods such as skip counting, repeated addition, and manipulating counting materials can also be used. Some students, realizing that the problem involves division and multiplication, may promptly ask for a calculator. After discovering how much the magician owes, your students might enjoy writing stories telling how he can earn the necessary funds.

After the prince and princess flew away from his kingdom, the King of Cashmere looked over the list of prisoners. He saw the magician's name.

"This man has been in jail long enough," said the king. "Let him go, but first make him pay for his meals."

The magician ate 3 meals a day in jail. He ate a total of 270 meals. How many days did he spend in jail?

The magician's food cost the king $5.00 a day. How much money does the magician owe for his meals?

The Ebony Horse

Getting Out of Jail

What about the magician? The King of Cashmere agrees to let him leave prison, but first he has to pay for all his meals.

The magician ate 3 meals a day in jail.

He ate a total of 270 meals.

How many days did he spend in jail?

The magician's food cost the king $5.00 a day.

How much money does he owe for food?

Guarding the Horse

Students must evaluate a considerable amount of information in order to solve this problem. The prince needs guards 7 days a week. At 24 hours a day, that's 168 hours. Each guard works only 28 hours a week. That means it takes 6 guards to protect the horse around the clock. Because the prince wants a two-guard team at all times, he will need 12 guards. Obtaining this result involves a great deal of multiplication and division. Of course, students can use addition and subtraction or trial-and-error instead.

Making a schedule for 12 guards covering 24 hours a day for 7 days is also a complicated business. Children will need a way to identify the guards before assigning hourly responsibilities. You may decide not to assign this part of the problem if you think it is too great a task for your students. Students can create many possible work schedule for the guards.

The prince, wanting to take proper care of the flying horse, built a special stable for this amazing animal. To protect the horse from danger, the prince insisted that two guards watch the stable at all times.

The guards in Persia work exactly 28 hours a week, no more, no less. Knowing this, the prince has to determine how many guards he needs for 'round-the-clock, two-guard protection. Help him solve this problem.

When you know how many guards the prince needs, help him create a one-week work schedule for the guards. You must consider these two important facts while figuring out this problem. First, a Persian guard works only 5 days a week. Second, the longest any guard can be on duty in a work day is 7 hours.

The Ebony Horse

Guarding the Horse

The prince built a stable for the flying horse. He insists that 2 guards watch the stable at all times.

Each guard works exactly 28 hours a week. How many guards does the prince need to protect the horse for one week?

Now create a one-week work schedule for the guards. Make sure that each guard works 5 days a week and that the longest time on duty for any guard is 7 hours a day.

The Ebony Horse

The Princess' Game

This game, played with four dice, is based on chance. It can go on for quite a while, giving students lots of addition and subtraction practice. You may want to review the meanings of odd and even numbers with your students before they play.

The princess missed many things about her old home. She especially missed playing her favorite dice game. No one in Persia knew this particular game. What did the princess do about this problem? She taught the prince and all his sisters and brothers how to play. You can play, too.

Here are the rules.

- Each player begins the game with a score of 100 points. The first player rolls four dice and then adds to find the sum of the numbers rolled.

- If the sum is an odd number, the player adds it to his or her score. If the sum is an even number, the player subtracts it from his or her score.

- Players take turns rolling the dice and adjusting their scores. A player whose next roll takes him or her over 200 points or below 0 is the winner!

A Tale from the United States

At the dawn of time, the world—the entire world—was covered by one never-ending ocean. Every manner of fish, swimming bird, and animal lived in the endless sea. Far above the water and hidden by clouds was Skyland. In those first days, all people dwelled in that distant place. They lived below the leafy canopy of a magnificent tree whose roots stretched to the four corners of the universe. Fruits, vegetables, and flowers hung down from the tree's branches. Whenever the Skypeople were hungry, they plucked handfuls of ripe, juicy treats off the limbs of their wondrous tree.

The chief of Skyland was a very old man married to a very young woman. One night the chief's wife dreamt that the tree, the life-giving tree of Skyland, was trembling and shaking. Suddenly it wiggled loose from its roots and began floating—floating freely throughout Skyland.

In the morning, the chief's wife related the dream to her husband. Instantly the chief realized that this was no ordinary dream. No, this was a power dream. Whatever the diffi-culty, whatever the consequences, the Skypeople had to turn his wife's vision into a reality. The tree, the great tree, must be uprooted.

And so the chief assembled all the Skypeople. He explained what needed to be done and the people understood. The young men circled the tree trunk. They pulled and pushed, shoved and rammed. They worked together and they worked separately, but they could not dislodge the tree. Finally the chief, the old, very old chief approached. He embraced the trunk, holding it tighter and tighter. Then, exerting one great effort, he ripped the tree from its roots. Unhindered, the tree floated serenely around Skyland.

Where the tree once stood, a gaping hole appeared in the clouds. The chief's wife approached the opening and looked through it. What did she see? Shimmering, glimmering, sparkling water. She was amazed, fascinated, mesmerized. Wanting a closer look, she bent further and further over the hole. To keep her balance, she grabbed onto branches of the great tree which was, at that very moment, floating overhead. But she stretched too far and her hand slipped. She fell through the hole and plunged toward the water.

Far below, the animals saw a great hole materialize in the sky far above their heads. They saw the chief's wife as they stared downward. And then they saw her tumble. At first, they were startled and then, worried.

"Look," exclaimed Crane, "the falling one does not have wings."

"She cannot fly," observed Loon.

"She will crash and die," mourned Duck.

"We can help her," two swans announced in unison.

Together the swans flew up high, high, and higher until they reached the chief's wife. After catching her, they held her aloft on their wide, powerful wings.

"Skywoman has strange feet," said Crane.

"They are not webbed," noted Loon.

"How can she swim without webbed feet?" wondered Duck.

"How can she live in the water?" asked Turtle.

Again the animals worried. True, the swans were strong, yes, very strong, but even they could not hold Skywoman forever. What would happen when they got too tired to fly?

Otter had an idea. "Yesterday, while swimming far under the surface of the water," she said, "I saw something new in the distance below. Earth—it was the earth. I believe that Skywoman could live on this earth, if we could get it for her."

The animals knew that Otter was very intelligent and so they listened carefully to her words. But there was still a difficulty— could any animal swim far enough, deep enough to get this earth for Skywoman?

Everyone considered this problem. Then Duck said, "I do not know if I can reach the earth, but I will try."

With that, he dove under the water. He swam and swam but could not reach bottom.

Next Beaver tried. She got a little further than Duck, but not far enough.

Loon made the attempt but did not succeed.

One by one the animals took turns, and one by one they failed to reach their goal. Finally Muskrat offered to try. Muskrat was a small animal, not as swift or strong as the others, but very determined to get some earth for Skywoman.

He started swimming. He forced himself lower and lower. His heart pounded and his legs ached. His lungs were ready to burst, but he kept swimming. Just as he could barely stand it another moment, he touched bottom and scooped up a pawful of oozing mud. Holding it as best as he could, he paddled back to the surface.

There, gasping for air, he showed the animals his prize.

"Where can we put this earth?" asked Loon.

"On my back," answered Turtle.

And so Muskrat placed his precious mud on Turtle's shell. Instantly Turtle started growing. He grew bigger and bigger and bigger until his back, covered with earth, turned into all the continents of the world.

The swans lowered Skywoman onto the new land. As her feet touched the ground, she opened her hands. Seeds fell from her fingers—seeds from the great tree—seeds that had remained in her palm when she let go of the tree's branches. Everywhere the seeds fell flowers, grass, bushes, ferns, and trees started growing.

This is how earth came to the watery world. And this is how life, flowering, blooming, glowing, growing life came to the earth.

Notes

About the Story

This is an Iroquois tale originating with the Onondaga tribe. Most societies have creation myths—stories that explain how the world came to be. The ancient Greek tales tell of Gaea, the Earth, who married Uranus, the Sky. In love, Gaea became Mother Earth who gave birth to all living things. The Bakuba tribe of Zaire believes that the god Mbombo ruled over darkness and water. One day he vomited and up came the sun, moon, and stars. He vomited again and up came trees, animals, and people. Norse myths tell of great fires that cooled the icy region of Nifheim, causing the creation of the first two creatures—an evil frost giant and a huge cow. You should be able to find versions of these and other creation stories to share with your students.

In addition to creation myths, you can find tales that explain all sorts of natural occurrences. You may discover stories that tell why volcanoes erupt, why the moon has phases, and why buzzards lack neck feathers (a legend that readers of *Afterwards,* grades 1 and 2 already know).

With your encouragement, students might try writing their own natural histories. Would anyone care to explain the origins of rainbows? camel's humps? shooting stars?

For this retelling, I used one version of the story.

> Bruchac, Joseph. *Native American Stories*. Colorado: Fulcrum Publishing, 1991, 5-9.

Your Thoughts

On Turtle's Back
Coming to Earth

Students must record the sums and differences resulting from repeatedly adding ten, then subtracting two. Students having difficulty keeping track of their findings may find it easier to work in pairs. Using base-ten materials, one member of the pair can manipulate the materials while the other records results. Alternatively, some students will realize that adding ten and subtracting two is the same as merely adding eight. These students will add or multiply to find that it took 13 days for 106 Skypeople to join Skywoman on the earth.

The chief and all the people of Skyland saw Skywoman fall through the hole. They saw her resting on the swans' wings. They saw her settle on the new earth. They watched while trees and flowers grew over the world.

"I would like to visit this new land," said one of the young men.

"Me, too," said a young woman, "but I do not want to fall through the skies."

And then the chief—the ancient chief—had an idea. He hung a long rope, very long rope, all the way from Skyland down to the new earth. The Skypeople could descend the rope and visit the earth. Then they could climb back up and return to Skyland. As soon as the rope was in place ten brave men and woman slid down to the world below. At the end of the day, two of them wanted to return to Skyland, and so they did.

The next day, ten more Skypeople descended the rope. Once again, at the end of the day, two of them returned to Skyland. Everyday began the same way—ten people climbed down to the earth, then at day's end, two climbed back up.

One bright afternoon Skywoman counted and discovered that there were 106 other people with her on the earth. How many days had it taken for all 106 Skypeople to join Skywoman?

Coming To Earth

The chief dropped a rope from
Skyland to the earth.

Every morning 10 Skypeople
climbed down the rope.

Every evening 2 Skypeople
returned to Skyland.

How many days did it take for
106 Skypeople to get to the earth?

On Turtle's Back

Generations

Students must keep track of several generations of Skywoman's offspring. Skywoman has 3 children, 9 grandchildren, 27 great grandchildren, and 81 great, great-grandchildren. Altogether, that's 120 people, two thirds (or 80) of whom are girls, one third (or 40) of whom are boys.

One way to solve this problem is to make tally marks. Start with three tallies at the top of a page to represent Skywoman's children. Then let three tallies branch off from each initial tally. This second row of marks represents Skywoman's grandchildren. Students can keep branching off for each generation. As they do this, they will discover a variety of patterns and a great many opportunities to multiply by three.

After living on the earth for a short time, Skywoman had children—three children—triplets—two girls and one boy. When her children were young, they loved listening to their mother tell the story of how she once lived in Skyland, how she fell from there, how the swans held her in the air, how the muskrat collected earth for her, and how the turtle came to hold up the world.

Years passed and Skywoman's children grew up. They became parents themselves. And the odd things is, they each had triplets—two girls and a boy. These girls and boys loved hearing the tale of their grandmother's adventures, and when they grew up, they had their own children—triplets—each of them had triplets—two girls and a boy. These were Skywoman's great grandchildren.

Each of Skywoman's great grandchildren had triplets—two girls and a boy. These were her great, great-grandchildren.

By now Skywoman was very, very old, but once a year she invited all her children, grandchildren, great grandchildren, and great, great-grandchildren to her home. On that special day, she told and then retold the story of her life for all to hear.

You know how many children Skywoman had. Now figure out how many grandchildren, great grandchildren, and great, great-grandchildren she had. How many people joined Skywoman on the special day she told her story? How many of them were girls? How many were boys?

On Turtle's Back

Generations

Skywoman had three children—triplets—two girls and one boy.

Skywoman's children had children of their own. They each had triplets—two girls and one boy. These were Skywoman's *grandchildren*.

Skywoman's grandchildren had children. They each had triplets—two girls and one boy. These were Skywoman's *great grandchildren*.

Skywoman's great grandchildren had children. They each had triplets—two girls and one boy. These were Skywoman's *great, great-grandchildren*.

Once a year, Skywoman invited all her children, grandchildren, great grandchildren, and great, great-grandchildren to her home. On that special day, she told and then retold the story of her life for all to hear.

You know how many children Skywoman had. Now figure out how many grandchildren she had.

How many great grandchildren did she have?

How many great, great-grandchildren did she have?

How many people joined Skywoman on the day she told her story?

How many of them were girls?

How many were boys?

A House for Skywoman

There are many parts to unravel in this problem. To begin, students must discover the number of blocks needed to build Skywoman's house. This takes two calculations, the first to find that each wall has 42 blocks, the second to learn that 252 blocks are needed for six walls. Students can multiply to get these results, but they can also use less formal approaches, like counting or adding. They must then find out how many days it takes to make the blocks. To do this, students start with 10 to which they add a string of numbers that increase by two—10 + 12 + 14 + 16 + ⋯ + 32 in order to reach the sum of 252. It takes twelve addends (12 days) to reach this sum. After solving this problem, you may have your students design a house of a different shape for Skywoman. (Of course, they must then determine the number of blocks needed for their constructions.)

Skywoman was now living on the earth, but she did not have a home. The animals could see that Skywoman needed shelter from wind and rain. So they decided to build her a house. They would use big clay blocks for walls and straw for the roof.

Some animals thought a rectangular house would be best. Others wanted one in the shape of a pentagon. Still others argued for a house shaped like a hexagon. They took a vote and the hexagonal house won.

The animals decided that each side of the house would be six blocks long and seven blocks high.

Everyone worked together to make the clay blocks. At first it was hard, slow work, but after practicing each day, it got easier.

On the first day, the animals made only 10 blocks; the second day, they made 12 blocks; the third day, they made 14. Each day they made two more blocks than they had made the day before.

At this rate, how many days did it take the animals to make enough blocks for Skywoman's house?

A House for Skywoman

The animals wanted to build a house for Skywoman.

They decided to create a house in the shape of a hexagon.

It will have clay blocks for walls and straw for the roof.

Each side will be 6 blocks long and 7 blocks high.

The animals worked together to make the blocks.

The first day, they made 10 blocks.

The second day, they made 12 blocks.

The third day, they made 14 blocks.

Each day the animals made two more blocks than they made the day before.

At this rate, how many days did it take to make enough blocks for Skywoman's house?

Party Food

For students to solve this challenging problem, they must work backwards and forwards to find out how much fruit is needed for Skywoman's party. It is probably best for students to act out the problem. They can draw or use counting materials to aid them. Along the way, they may stumble upon some helpful patterns.

The correct solution is 30 apples cut into 240 slices, 24 bananas cut into 120 slices, 20 oranges cut into 80 slices, 8 cantaloupes cut into 80 slices, and 20 plums cut into 40 slices.

Skywoman wanted to thank the animals for their help, so she decided to have a party—a fruit party. None of the animals had ever eaten fruit before, and Skywoman thought it would be delightful to present them with a selection of sweet juicy goodies. Carefully, she picked fruit from all the new trees, bushes, and vines. Then she cut each piece of fruit into slices. She cut each apple into 8 slices, each banana into 5 slices, each orange into 4 slices, each cantaloupe into 10 slices, and each plum into 2 slices.

Skywoman used banana leaves for plates. On each leaf she put the same amount of fruit: 6 apple slices, 3 banana slices, 2 orange slices, 2 cantaloupe slices, and 1 slice of plum.

After filling 40 leaves, she did not have any fruit left over. Figure out how many apples, bananas, oranges, cantaloupes, and plums she used for the party.

Party Food

Skywoman decided to have a party for the animals.

She picked lots of fruit and then cut each piece of fruit into slices.

She cut each apple into 8 slices.

She cut each banana into 5 slices.

She cut each orange into 4 slices.

She cut each cantaloupe into 10 slices.

She cut each plum into 2 slices.

Skywoman used banana leaves for plates. On each leaf she put:

- 6 apple slices
- 3 banana slices
- 2 orange slices
- 2 cantaloupe slices
- 1 plum slice

After filling 40 leaves, she had no fruit left. Figure out how many apples, bananas, oranges, cantaloupes, and plums she used for the party.

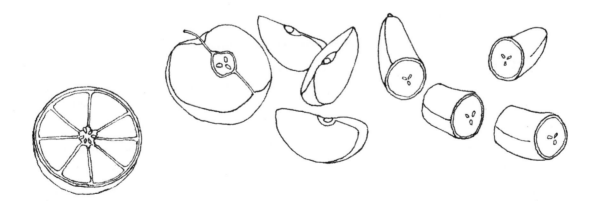

The Wonderful Healing Leaves

A Jewish Tale from Iraqi Kurdistan

Long ago and far away a powerful king had three exquisitely beautiful daughters. The king wanted his children to marry wealthy, handsome princes, and the two older girls did exactly that. The youngest princess, Sofia, had her own ideas. She wanted to marry Arik, a young man who, although neither rich nor good-looking, was kind and wise. Sofia loved him with all her heart.

The king refused to allow this marriage, but Sofia did not listen. She wed Arik anyway and when her father found out, he banished her from the palace. So the princess went to live in her husband's home, a small hut at the very end of town.

Shortly after Sofia and Arik's wedding, a strange and terrible thing happened. The king lost his eyesight. Yes, it is true, he was completely blind. From near and far, physicians came to the palace, but none of them had a remedy for the king's blindness. One day a doctor arrived from a very distant land. After examining the king's eyes, he made a solemn announcement.

"Your eyesight can be restored, Sire, but only if you rub your eyes with the healing leaves which sprout from a magic tree that grows in the Land of No Return."

"Wonderful, wonderful," said the king, "obtain those leaves for me, and I shall pay you handsomely."

"That I cannot do," said the doctor, "and neither can anyone else. For no one has ever survived a trip to the Land of No Return and come home again."

The king was not discouraged by these words. Turning to the husbands of his two eldest daughters he said, "My sons, you shall travel to the Land of No Return, gather the healing leaves, and bring them to me. When I regain my vision, I will give each of you one third of my kingdom. Fail in this mission, however—come home without the leaves—and my royal executioner will hang you both."

Having made this pronouncement, the king handed each prince a pouch full of gold, mounted them on the fastest horses in the royal stable, and sent them on their quest.

After many days, the two princes reached a small village that bordered the Land of No Return. They stopped at an inn and asked the innkeeper about the tree with healing leaves.

"Everyone wants those leaves," said the innkeeper, "for they cure all maladies. But no one can get to them. The tree grows somewhere in a castle that belongs to a mighty and magical queen. A viper and a dragon guard her home. These beasts kill anyone who dares approach the castle gates."

The innkeeper's words distressed the princes. "We cannot risk going after the healing leaves," said one.

"Certainly not, but we cannot go home without them," said the other.

"What shall we do?" asked the first.

"This inn seems a pleasant place," said his companion. "We might as well stay here."

And that is just what they did. They stayed

at the inn, using the king's gold to pay for the best rooms and the tastiest foods.

A few days after the princes left the palace, Sofia learned of their search. Immediately she went to her father.

"You must allow Arik to seek the healing leaves. He is so kind and wise, he will surely succeed where others have failed," she said.

"Fine, let him go," grumbled the king. "But if he returns without the leaves, I will tell the royal executioner to hang him."

"And," added Sofia, "when he returns with the leaves, he will get one third of your kingdom."

The king agreed, but he gave Arik his oldest, slowest horse and, without providing him with so much as a copper coin, sent him on his way.

Eventually Arik arrived at the inn near the Land of No Return. There he saw the two princes eating a hearty meal. The royal pair, however, were too preoccupied with their food to notice a tired and dusty traveler.

Approaching the innkeeper, Arik asked for help in finding the magic tree that sprouts healing leaves. The innkeeper told him about the queen, her castle, the viper, and the dragon. Arik was not a bit frightened by this tale. Indeed, the young man begged to know the fastest route to that terrifying place.

"In the valley below," said the innkeeper, "there lives a giant. He alone knows the way to the castle. If you are as brave as you seem, visit him. But I warn you, he is a brutal monster and human flesh is his favorite meal."

Undeterred, Arik got on his horse and headed for the valley. When he saw the giant's house he did not hesitate, but rode straight through the open door. Once inside,

Arik saw a colossal being sitting on the floor stripping bark off the trunk of an oak tree. When the giant completed this job, he stuck the trunk in his mouth and used it as a toothpick.

"Hello, Sir," shouted Arik upward toward the giant's massive head. "Tell me, how can I get the healing leaves that grow in the Land of No Return?"

The giant was astonished to hear a strange voice in his house. He was even more amazed to see a tiny man standing resolutely before him.

"You are smaller than my thumbnail," roared the towering monster, "but in entering my home you have shown a giant's bravery. I must reward such courage. Yes, little man, I will answer your question. Listen carefully and remember all I say.

"Ride for seven days along Queen's Road. As the sun sets on the seventh day, the road will seem to reach a dead end. Do not be discouraged. Instead, call out, 'What a beautiful path!' and the road will reappear. In the distance you will see a castle. Approach with caution, for a 100-foot viper twists and turns about the castle gate. Wait patiently until the beast's eyes open. Only then does he sleep. Slide quickly through the bars of the gate while saying, 'What a lovely day for a walk!' Once you have said these words, the viper will not wake up.

"You must now find the queen's bedroom. It is the one guarded by a dragon. Wait until the dragon's eyes open, then enter the room while saying, 'This must be a dream.' Hearing this phrase, the queen will fall asleep. Next to her bed, you will see the tree with the healing leaves. Pick all the leaves from every branch—

The Wonderful Healing Leaves

even the tiniest buds. Put some into your leather pouch and the rest into your pockets. That done, remove your iron wedding ring and exchange it for the golden band on the queen's left hand. When her ring rests on your finger, you will be free to leave the Land of No Return."

Arik did exactly as the giant advised. He traveled the Queen's Road, entered her castle, gathered her healing leaves, took her golden ring, and returned safely to the inn.

When he arrived, the two princes were sitting at a long table gobbling huge bowls of lamb stew. Between mouthfuls, they saw Arik enter and request a room. At first they were simply surprised to see their brother-in-law, but then they noticed the bright green leaves poking out from the corners of his worn-out leather pouch.

"That irritating fellow has found the healing leaves," moaned one prince.

"Yes, and he will carry them to the king and take our reward," whined the other.

"Unless," they whispered in unison, "we do something to stop him."

Then and there the princes concocted a cruel plan. Late that night they crept into Arik's room, tied him up with a strong rope, shoved him into a burlap bag, carried him far into the woods, and left him there hidden in a deep hole under a gnarled tree.

"If fate declares that wolves devour him, it won't be our fault," snickered the princes as they returned to the inn.

The next morning, carrying Arik's pouch of healing leaves, they started home.

Naturally, the king was overjoyed when the princes returned with the leaves. Giggling in anticipation, he grabbed a handful and rubbed them over his eyes. As soon as he did, his vision returned. He could see everything—even the ant in his garden and the tiniest mosquito buzzing about his courtyard. That night the king celebrated his renewed sight by throwing a huge party. At the end of the evening, he rewarded the princes with vast lands and trunks full of gold.

What about Arik? What happened to him? He struggled for a night and a day before escaping from the burlap bag. Returning to the inn, he mounted his old horse and headed home. "I don't have a pouch full of leaves," he said to himself, "but I do have pockets full."

Several days later, Arik arrived at the hut he shared with Sofia. After hugging his dear wife, he eagerly recounted his adventures and showed her the healing leaves. Tears in her eyes, Sofia explained that the princes had already cured her father.

Of course, Arik and Sofia knew how the princes obtained the magic leaves, but they also knew that no one would ever believe their story.

In time though, the king learned that things are not always as they seem. You see, back in the Land of No Return, the queen awoke from her long, long sleep. Immediately, she realized that her leaves had vanished and that her golden ring was gone. How anyone could invade her house she did not know, but she vowed to find the culprit.

She climbed onto her magic carpet and started flying. She soared here and there—far and wide. Eventually she heard stories of a king who had been blind but, thanks to a miracle cure, could now see.

"Ah," said the queen, "my leaves!"

Rushing to the king's palace, the queen landed her carpet before a startled monarch.

"How, Sir," the queen asked, "did you cure your blindness? Tell me now and tell me honestly or I will order my dragon to destroy your palace."

The princes, who were standing alongside the king, held up Arik's pouch.

"We found this bag of leaves in the forest," said one, trembling in his boots.

"Hanging from a tree," added the other, shaking in his shoes.

"Your words are false," intoned the queen. "I shall send for my dragon."

Overcome with terror, the princes confessed everything.

When the queen heard the truth she demanded to see Arik. Moments later, the young man entered the palace, told his story, and then without flinching, held up a handful of leaves and the golden ring.

"You are courageous as well as honest," said the queen. "Therefore, I will not punish you. Indeed, you may keep the ring as a token of my esteem." Then, as suddenly as she had appeared, the queen departed.

As soon as she left, the king banished the cowardly princes from his kingdom and gave all their land and wealth to brave Arik and loyal Sofia. From that day on, Arik and Sofia lived happily ever after.

Notes

About the Story

I used two sources in retelling this story:

Lester, Julius. *How Many Spots Does A Leopard Have? And Other Tales.*
New York: Scholastic Inc., 1989, pages 59–68.

Schwartz, Howard. *Elijah's Violin and Other Jewish Fairy Tales.* New
York: Harper and Row, 1983, pages 163–168.

Your Thoughts

The Wonderful Healing Leaves

Wishes from the Ring

To find the days on which the ring grants wishes, children must discern a subtle pattern of calendar dates. The pattern is as follows: skip *one* day, skip *two*, skip *three*, skip *two*, skip *one*, and so on. Trying to find logic in a span of numbers can be very difficult, and that's why many children find complex pattern-hunting frustrating. But then, learning to work through frustration is an important skill. You can discover a lot about how well children understand patterns by observing the patterns they create for the magic ring.

In time, Arik realized that his golden ring grants wishes, but that it doesn't grant them every day. Indeed, it grants one wish on some days and no wishes at all on others.

For a long time, Arik made daily wishes hoping the ring will work, but he worried that so much wishing will use up the ring's powers.

Then he had a good thought—perhaps there is a pattern to the days the ring grants wishes. Hoping so, Arik decided to keep track of the wish-granting days on a calendar. Throughout all of April and the first week in May, he wrote a big X on the calendar every day he got his wish.

Then, on May 8, while studying the calendar, Arik saw a pattern. Can you see it, too? When you do, continue the pattern by recording X's on other days in May on which the ring grants wishes.

One day, the queen came to visit Arik. She said, "I want the ring to follow a new pattern of wish-granting days in June and July. You can invent it yourself, but do not ask the ring to grant more than eight wishes over 30 days. If you succeed, the ring will grant your wishes. If you do not, the ring will lose its power."

After that, the queen flew away on her carpet. Help Arik invent a pattern of days that meets the queen's demand.

Wishes from the Ring – 1

Here are Arik's calendars for April and May.

Write more X's on the May calendar to complete the pattern of wish-granting days.

April

S	M	T	W	T	F	S
1	2 X	3	4	5 X	6	7
8	9 X	10	11	12 X	13	14 X
15	16	17 X	18	19	20	21 X
22	23	24 X	25	26 X	27	28
29 X	30					

May

S	M	T	W	T	F	S
		1	2	3 X	4	5
6 X	7	8 X	9	10	11	12
13	14	15	16	17	18	19
20	21	22	23	24	25	26
27	28	29	30	31		

Wishes from the Ring – 2

Now make up a new pattern of wish-granting days.

Write X's on the June and July calendars
to show your pattern.

June						
S	M	T	W	T	F	S
					1	2
3	4	5	6	7	8	9
10	11	12	13	14	15	16
17	18	19	20	21	22	23
24	25	26	27	28	29	30

July						
S	M	T	W	T	F	S
1	2	3	4	5	6	7
8	9	10	11	12	13	14
15	16	17	18	19	20	21
22	23	24	25	26	27	28
29	30	31				

The Wonderful Healing Leaves

A Giant Dinner

The princes need 60 servings of each recipe for the giant's dinner. They must increase each of the ingredients in the original lamb stew recipe 10 times and each of the ingredients in the original cake recipe 6 times. They can do this by multiplying or adding. Here are the "giant-size" recipes.

<u>Lamb Stew</u>: 30 tbs butter, 35 lb lamb, 12½ tsp salt, 1¼ tsp pepper, 5 lb small white onions, 6⅔ lb mushrooms, 10 cloves garlic, 23⅓ tbs flour, 30 sprigs parsley, 5 bay leaves, 1¼ tsp dill

<u>Chocolate Cake</u>: 12 c cake flour, 10½ c sugar, 4½ c cocoa, 7½ tsp baking soda, 3 tsp baking powder, 6 tsp salt, 4½ c shortening, 7½ c milk, 6 tsp vanilla, 18 eggs

What happened to the two cowardly princes? Banished and in disgrace, they traveled here and there. Eventually they wandered into the giant's valley and found themselves at his front door. And then things got worse—the giant caught them up in the palm of his hand.

"You two should make a fine dinner," he laughed.

"Please Mr. Giant, do not eat us. We are not tasty," cried one of the princes.

"But we can prepare a delightful dinner for you," added the other.

"Put us down, and we will start cooking," pleaded the first.

The giant, tired of eating raw food, agreed to free the little men.

So the princes had to cook. They decided to make lamb stew and chocolate cake. But they still had a problem. You see, their stew recipe is meant to feed 6 people and their cake will feed 10. The giant likes to eat 60 servings of everything. Clearly, in order to have enough food, the princes must increase the amount of each ingredient.

Here are the princes' recipes. Figure out how much of each ingredient they need to cook for the giant.

A Giant Dinner

The giant eats enough for 60 people. Increase the amount of each ingredient in the lamb stew and chocolate cake recipes so that there will be enough to feed the giant.

<u>Lamb Stew</u>—This recipe serves 6.

3 tablespoons butter

$\frac{1}{8}$ teaspoon pepper

$\frac{2}{3}$ pound mushrooms

$2\frac{1}{3}$ tablespoon flour

$\frac{1}{8}$ teaspoon chopped fresh dill

$3\frac{1}{2}$ pounds lamb cut into chunks

$1\frac{1}{4}$ teaspoon salt

$\frac{1}{2}$ pound small white onions

1 clove garlic

3 sprigs parsley

$\frac{1}{2}$ bay leaf

<u>Chocolate Cake</u>—This recipe serves 10.

2 cups cake flour

$\frac{3}{4}$ cup cocoa

$\frac{1}{2}$ teaspoon baking powder

$\frac{3}{4}$ cup shortening

1 teaspoon vanilla extract

$1\frac{3}{4}$ cups sugar

$1\frac{1}{4}$ teaspoon baking soda

1 teaspoon salt

$1\frac{1}{4}$ cup milk

3 eggs

The Wonderful Healing Leaves

How Many Cures?

Logical, organized thinking and list making will help most children discover each combination of patients that Arik can cure with 45 g of medicine. Some children, however, may arrive at their solutions by trial-and-error alone. Here are all nine combinations.

Combination	Babies	Children	Adults
1	9	1	1
2	7	2	1
3	5	3	1
4	3	4	1
5	1	5	1
6	5	1	2
7	3	2	2
8	1	3	2
9	1	1	3

Arik plants a few leaves in the palace garden hoping to grow a magic tree that will sprout healing leaves. It works. Each week he picks all the new leaves and pounds them into a delicate powder. In this way he gets 45 g of medicine—medicine that can cure any disease.

After some experimenting, Arik discovers that the treatment for a baby requires exactly 3 g of medicine. To treat a child, he needs 6 g of medicine. For an adult, he needs 12 g.

Every week Arik uses all 45 g of medicine. He always treats at least one baby, one child, and one adult.

There are nine possible ways he can cure patients each week. For instance, in one week he may cure 3 babies, 2 children, and 2 adults. What are the eight other ways?

How Many Cures?

Every week Arik makes 45 g of medicine.

- To cure a baby, he uses 3 g.
- To cure a child, he uses 6 g.
- To cure an adult, he uses 12 g.

Every week he uses all 45 g and cures at least one baby, one child, and one adult.

Following this plan, there are nine possible ways he can cure patients each week. For instance, he can cure 2 adults, 2 children, and 3 babies. Find the eight other ways.

12g 12g 6g 6g 3g 3g 3g

More Healing Leaves

This is a multi-step division problem. Students must divide the total number of leaves Arik picked (600) by the number of weeks over which he picked them (6) and then divide the quotient (100) by the number of tree branches (20) to find that 5 leaves had sprouted on each branch.

To solve the problem children may use the strategies of working backwards or trial-and-error. Making drawings and manipulating counting materials will prove useful, too.

One morning Arik took a handful of precious healing leaves and gently, carefully, planted them in one place in the palace garden. This, it turns out, was an excellent idea. Within one hour, a tree appeared in that very spot. It was a magic tree and it grew healing leaves.

Only Arik could pluck the leaves from this tree. If anyone else tried, the leaves withered instantly. The tree had exactly 20 branches. Each and every branch produced an identical number of leaves. Every Monday morning, Arik picked them all, down to the tiniest bud.

By the end of six weeks, Arik had collected a total of 600 leaves. Your job is to figure out how many leaves sprouted each week on each branch of the magic tree.

More Healing Leaves

One day Arik planted some of his healing leaves in one place in the palace garden. Within one hour, a tree appeared in that spot. It was a magic tree that grew healing leaves.

The tree had exactly 20 branches and each branch produced an identical number of leaves.

Every Monday morning Arik picked all the leaves on the tree. By the end of six weeks, Arik had 600 leaves. Can you figure out how many leaves sprouted each week on each branch of the magic tree?

One, two, three, four, five, six...
